Simplified Metal Works
（金工操作）

战 盈　王朝辉　编

西北工业大学出版社

图书在版编目(CIP)数据

金工操作＝Simplified Metal Works:英文/战盈,王朝辉编.—西安:西北工业大学出版社,2014.9

ISBN 978-7-5612-4160-8

Ⅰ.①金… Ⅱ.①战… ②王… Ⅲ.①金属加工—教材—英文 Ⅳ.①TG

中国版本图书馆 CIP 数据核字(2014)第 220228 号

出版发行:	西北工业大学出版社
通信地址:	西安市友谊西路 127 号　邮编:710072
电　话:	(029)88493844　88491757
网　址:	www.nwpup.com
印　者:	陕西兴平市博闻印务有限公司
开　本:	787 mm×1 092 mm　1/16
印　张:	13.375
字　数:	320 千字
版　次:	2014 年 11 月第 1 版　2014 年 11 月第 1 次印刷
定　价:	30.00 元

Foreword

To teach any subject, especially in higher learning, textbooks are usually a necessity. For the study of chemistry, physics, mathematics, etc. in the classroom and the laboratory, there are well-written textbooks to guide the student progressively through elementary and advanced principles. It is unimaginable even for Chinese students to study any course without a good textbook. It is more so for students study in a foreign country, where the language poses a great barrier.

However, in this modern era of machining, international students and teachers of metal works often found themselves caught in a helpless situation: the lack of English metal work textbooks makes it impossible for such lessons to go on. On the one hand, the teacher doesn't have a suitable English textbook to lecture and impart their knowledge; on the other hand, international students have to go with Chinese textbooks which they might not know even a single Chinese word.

To meet the urgent demand for such a textbook, the author compiled a Simplified English Metal Works with a view to offering a much needed assistance to those who are handicapped by textbooks.

Owing to the limited space and content, this book is by no means comprehensive, which an experienced reader may find wanting in many ways, however, if the book can give international students a glimpse of what metal works pertains to, of its scope and application within 1 or 2 weeks of workshop practice, then, this book can be deemed as satisfying for the purpose.

Given the academic level of international students and their teachers' language skill, as well as the class requirement, this book will focus more on operations instead of theoretical pedagogy.

The authors
May 2014

Contents

Chapter 1　Basic Properties of Metals 1

  1.1　Basic Properties of Most Commonly Used Metals in Metal Practice 1
  1.2　Heat Treatment of Metals 5
  Exercises 6

Chapter 2　Technical Drawings 7

  2.1　Metal Shapes and Description 7
  2.2　Technical Drawing 8
  2.3　Measuring Equipment 12
  Exercises 19

Chapter 3　Casting 20

  3.1　Brief Introduction 20
  3.2　Basic Concepts 20
  3.3　Pattern 21
  3.4　Core and Core Prints 26
  3.5　Binders 28
  3.6　Molding Process and Materials 29
  3.7　Kinds of Molding Sand 30
  3.8　Classification Based on the Mold Material 32
  3.9　Steps Involved in Making a Sand Mold 34
  Exercises 36

Chapter 4　Forging 37

  4.1　Brief Introduction 37
  4.2　Forging Processes 37
  4.3　Classification of Forging Operations 41
  4.4　Die Design 43
  4.5　Type of Forging Machines 44
  Exercises 46

Chapter 5　Introduction to Welding 47

  5.1　Brief Introduction 47

5.2	Welding Process	47
5.3	Filler Metals	51
5.4	Fluxes	51
5.5	Weld Joints	52
5.6	Types of Welds	56
5.7	Welded Joint Design	60
5.8	Welding Positions	64
5.9	Expansion and Contraction	66
5.10	Welding Procedures	69
5.11	Safety	73
	Exercises	77

Chapter 6 Lathes ... 78

6.1	Brief Introduction	78
6.2	Types of Lathes	79
6.3	Engine Lathes and Classification	79
6.4	Lathe Components	80
6.5	Care and Maintenance of Lathes	82
6.6	Tools and Equipment	83
6.7	Lathe Attachments	94
6.8	Tools Necessary for Lathe Work	96
6.9	Basic Lathe Operations	98
	Exercises	102

Chapter 7 Planing Machines ... 103

7.1	Type of Planing Machines	103
7.2	Broaching	109
	Exercises	110

Chapter 8 Milling and Milling Machines ... 112

8.1	Introduction	112
8.2	Milling Machines	112
8.3	Major Components of Milling Machines	114
8.4	Milling Machine Accessories and Attachments	118
8.5	Mounting and Indexing Work	122
8.6	Milling Machine Operations	130
8.7	Milling Machine Adjustments	143
8.8	Milling Cutters	147
	Exercises	157

Chapter 9 Benchwork Tools, Drilling, Cutting, Sharpening ... 158

   9.1 Brief Introduction ... 158
   9.2 Benchwork Tools ... 158
   9.3 Work Bench ... 158
   9.4 Bench Vise ... 159
   9.5 Hand Hacksaw ... 159
   9.6 Chisel Tools ... 161
   9.7 Files ... 162
   9.8 Hammer ... 164
   9.9 Metal Cutting ... 164
   9.10 Drilling ... 166
   9.11 Cutting Threads with Tap & Dies ... 171
   9.12 Sharpening Tools ... 175
   Exercises ... 177

Chapter 10 A Brief Introduction to Modern Machining ... 178

   10.1 Brief Introduction ... 178
   10.2 A Short History of Modern Machining ... 178
   10.3 Key Developments ... 179
   10.4 Advantages of NC ... 180
   10.5 CNC Machining ... 180
   10.6 Advantages of CNC ... 181
   10.7 Classification of NC Machines ... 182
   10.8 CNC Programming ... 186
   Exercises ... 204

Chapter 1 Basic Properties of Metals

Teaching Objectives

Students are supposed to know the general properties of some most commonly used metals such as cast iron, machine steel, tool steel, high speed steel, brass and copper, and understand theheat treatment of them.

1.1 Basic Properties of Most Commonly Used Metals in Metal Practice

Of primary focus in metal works are such general properties of metals and their alloys as hardness, brittleness, malleability, ductility, elasticity, toughness, density, fusibility, conductivity, contraction and expansion. Such properties form the fundamental theoretical basis for further discussion of metal works. The following table lists some general information of the metals we commonly employed in daily metal works (Table 1.1).

Table 1.1　Metal Working Metals

Metal	Carbon Content/(%)	Appearance	Uses
Cast Iron (C. I.)	2.5 to 3.5	Grey, rough sandy surface	Parts of machines, such as lathe beds, water pump pitcher type, etc.
Machine Steel (M. S.)	0.10 to 0.30	Black, scaly surface	Bolts, rivets, nuts, machine parts
Cold Rolled (C. R. S.)	0.10 to 0.30	Dull silver, smooth surface	Shafting, bolts, screws, nuts
Tool Steel (T. S.)	0.60 to 1.5	Black, glossy	Drills, taps, dies, tools
High Speed Steel (H. S. S.)	Alloy Steel	Black, glossy	Dies, taps, tools, drills, toolbits
Brass		Yellow (various shades), rough if cast, smooth if rolled	Bushings pump parts, ornamental work
Copper		Red-brown, rough if cast, smooth if rolled	Soldering irons, electric wire, water pipes

To better understand the uses of them, the following metal properties are often described.

1.1.1 Melting Point

The melting point is the temperature at which a material starts to melt (Table 1.2).

Table 1.2　Melting Point of Common Metals　(℃)

Ferrous	1 536
Copper	1 083
Lead	327
Aluminum	658
Tin	232
Tungsten	3 387

1.1.2 Electrical Conductivity

This feature describes the ability of a metal to conduct electricity (Table 1.3).

Table 1.3　Electrical Conductivity of Common Metals

Copper	100%
Silver	106%
Lead	8%
Aluminum	62%
Ferrous	17%
Zinc	29%

1.1.3 Density

Density gives the quotient of mass and volume of a body (Table 1.4).

Table 1.4　Density of Common Metals　(kg/m^3)

Water	1.00
Copper	8.90
Lead	11.30
Aluminum	2.70
Steel	7.85
Tungsten	19.27

1.1.4 Thermal longitudinal Expansion

There is a coefficient describe such properties, with the length of 1 meter at a change of temperature of 1 degree Celsius (Fig. 1.1).

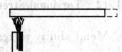
Fig. 1.1

1.1.5 Elasticity

Concerns metal's ability to return to its former shape after distortion. Heat – treated springs are good examples.

1.1.6 Ductility

Is the ability of a metal to be deformed for good without breaking. Examples may include copper and machine steel, which can be drawn into wire, are ductile (Fig. 1.2).

Fig. 1.2

1.1.7 Tensile Strength

Metal ability to resist fracture under tensile load (Fig. 1.3).

Fig. 1.3

1.1.8 Compressive Strength

The ability to withstand heavy compress load (Fig. 1.4).

Fig. 1.4

1.1.9 Brittleness

The property of a metal that allows no permanent distortion before breaking. Cast iron is brittle, which breaks rather than bends under shock or impact (Fig. 1.5).

1.1.10 Toughness

The opposite of brittleness, the metals ability to withstand shock or impact (Fig. 1.5).

Fig. 1.5

1.1.11 Shear Strength

Metal ability to resist fracture under shear load (Fig. 1.6).

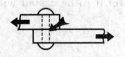
Fig. 1.6

1.1.12 Torsional Strength

Metal ability to resist torsional force (Fig. 1.7).

Fig. 1.7

1.1.13 Flexural Strength

The ability to resist under flexural force (Fig. 1.8).

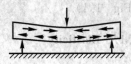
Fig. 1.8

1.1.14 Collapsing Stress

The metal ability to resist axial directed force (Fig. 1.9).

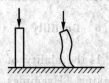
Fig. 1.9

1.1.15 Hardness

The metal ability to withstand abrasion or penetration (Fig. 1.10).

1.1.16 Weldability

The ability to weld 2 metals together. Weldability depends on the content of carbon. Steels with a content of max 0.22% are more or less weldable.

Fig. 1.10

1.1.17 Machinability

Indicates how easy materials can be machined (Fig. 1.11).

Fig. 1.11

1.1.18 Malleability

Is the property of metals that allows it to be hammered or rolled into other sides and shapes (Fig. 1.12).

Fig. 1.12

1.1.19 Castability

Is the property of metals that allows it to be molten and after it to be casted without any pores (Fig. 1.13).

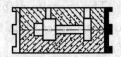
Fig. 1.13

1.1.20 Hardenability

Is the property of iron metals that allows it to increase the hardening through structural transformations (Fig. 1.14).

Fig. 1.14

1.2 Heat Treatment of Metals

Sometimes, the metals and alloys we are working with may not possess all the desired properties. To make them what we want, alloying and heat treatment are often used to improve the material properties.

By way of heat treatment, the microstructures of materials are modified and transformed, which lead to changes of mechanical properties, like strength, ductility, toughness, hardness and wear resistance. The purpose of heat treatment is to prepare the material for improved manufacturability

Heat treatment can be classified into the following types: hardening, annealing, normalizing, tempering and surface hardening.

1.2.1 Hardening

By heating up a metal or an alloy to a certain temperature and then cooling it rapidly, we can add strength and hardness to it.

In this process, take steel for example, steel is heated and held there until its carbon is dissolved, and then cooled rapidly in a way the carbon does not get sufficient time to escape and get dissipated in the lattice structure. This helps in locking the dislocation movements when stresses are applied.

1.2.2 Quenching

Sometimes we can cool hot metal rapidly by immersing it in brine (salt water), water, oil, molten salt, air or gas which in turn resulted in residual stresses (sometimes cracks). Residual stresses are removed by another process called annealing.

1.2.3 Annealing

During annealing, the steel will be heated to 780 – 930℃, and be kept at the temperature for required period of time, then to be cooled slowly. The cooling rate is around 10℃ per hour. The process has to be carried out in a controlled atmosphere of inert gas, to avoid oxidation. It is used to achieve ductility in hardened steels. Annealing can reduce hardness and remove residual stresses, improve toughness, restore ductility, and alter various mechanical, electrical or magnetic properties through refinement of grains.

1.2.4 Normalizing

During this process, the material will be heated above austenitic phase (1 100℃) and then cooled in air. It is similar to annealing but different in cooling methods. The air cooling speeds up the cooling process but leads to enhanced hardness and less ductility.

Normalizing is less expensive than annealing. Different sections of the workpiece will be cooled at different speed, hence the variation in properties.

The selection of which heat treatment is strongly influenced by the carbon content.

1.2.5 Tempering

During this process, the steel is heated to 350 – 400℃, and be kept there for about 1 hour and then cooled slowly at prescribed rate. Tempering is used to reduce brittleness and increase ductility of hardened steel like martensite (very hard and brittle). Tempering can also relieve stresses, hardness, strength and wear resistance marginally in martensite structure.

Exercises

1. How many metal properties can you remember? Can you name a few?
2. What is cast iron?

Chapter 2 Technical Drawings

Teaching Objectives

Students are required to master the basic knowledge of the dimensional description of various shapes and the skills needed, and learn how to do the technical drawing properly.

2.1 Metal Shapes and Description

Due to the wide variety of works in a metal shop, metals are manufactured in various shapes and sizes. Needless to say, there are proper methods for specifying the sizes and dimensions of metals. The following shapes are very common for metal jobs, and each comes with a way to describe them (Fig. 2.1).

Flat-bar: Thickness×Width×Length

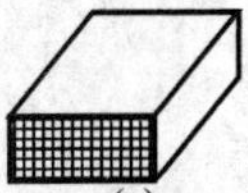

Round-bar: Diameter×Length

Square-bar: Width×Length

Angle-bar: Thickness×Width×Length

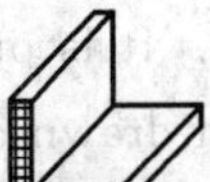

Hexagon-bar: Diameter×Length (or Distance Across Flats×Length)

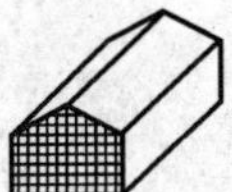

Pipe: Diameter×Schedule×Length #20 is thinner than #40

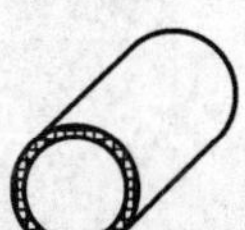

Fig. 2.1

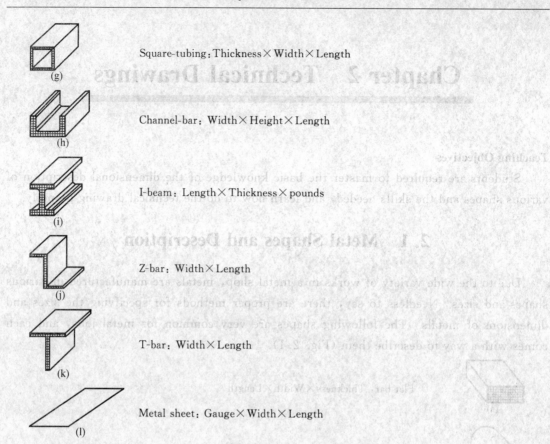

Cout. Fig. 2.1

2.2 Technical Drawing

2.2.1 Scale Size

Scale refers to the ratio of the drawing size to the actual size of the part. It is often necessary to enlarge small parts for clarity and details. Large objects are often drawn at a reduced scale in order to put down necessary information in a convenient piece of paper. The dimensions on the drawing give the correct size of the part required (Fig. 2.2).

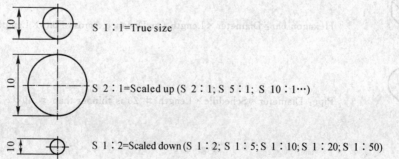

Fig. 2.2

2.2.2 Types of Lines

Technical drawings come with convention and standard to specify precisely what is required. The examples below is the so called "alphabet of lines" (Table 2.1).

Table 2.1 Types of Lines

	Type	Use
～～～	Free-hand line	Sketches; break line
▬▬▬▬	Object line	Indicate the visible form or edges of an object
▬▬▬▬	Thin unbroken line	Shading line, thread line, diagonal line
- - - - -	Hidden line	Indicate hidden contours of an object
— · — · —	Center line	Indicate centers of holes, cylindrical objects or other sections
←— 50 —→	Dimension line	Indicate dimensions of an object
↑- - - - -↑	Cutting-plane line	Show imagined section
▨▨▨	Cross-section line	Show surfaces exposed when a section is cut

2.2.3 Basic Rules

Dimensionsare entered in millimeters (Fig 2.3 – 2.9).

Dimension lines must have a distance of about 10 mm from the object edge and 7 mm from parallel dimension lines. The dimensions should be placed above the dimension lines and should be staggered. Dimensions must be put below or the right of the object edge. For small dimensions the arrows are placed outside.

Symmetrical workpieces are dimensioned symmetrical to the center line which extends 2 – 3 mm beyond the object edge.

Simple workpieces are usually drawn in front elevation.

If the area of a circle appears as a straight line, the diameter symbol must be placed in front of the dimension figure. If the circle is shown in the elevation, then the symbol is not necessary.

A radius is symbolized by R and has only 1 dimension arrow at the circumference. The center point is fixed by the crossing of center lines.

Concealed edges are drawn as dash lines. If visible and hidden edges coincide, the visible edges are drawn.

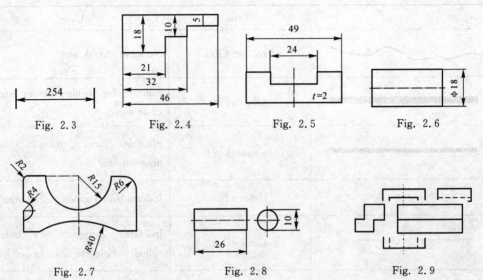

Fig. 2.3　　　Fig. 2.4　　　Fig. 2.5　　　Fig. 2.6

Fig. 2.7　　　Fig. 2.8　　　Fig. 2.9

Section views are used to show the interior form of an object. The section areas are shaded, not the hollow spaces. The shading lines are thin unbroken lines, which are angled at 45 degree to the center line or angled to the base edge.

In order to insert dimension figures the shading has to bebroken (Fig. 2.10 – 2.15).

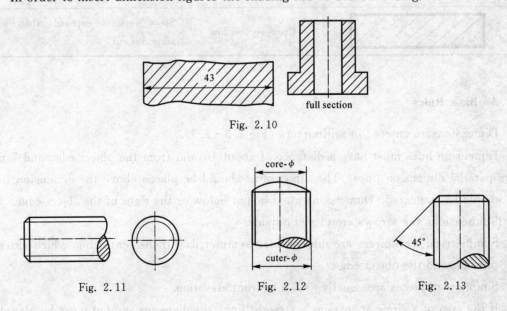

Fig. 2.10

Fig. 2.11　　　Fig. 2.12　　　Fig. 2.13

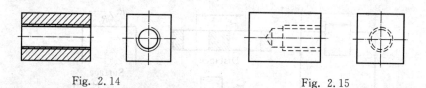

Fig. 2.14   Fig. 2.15

The outer diameter of a bolt thread is drawn as an object line, the core diameter as a thin unbroken line. The distance between the thick and thin lines represents the thread diameter.

Looking in direction of the shaft end the core diameter appears as a three-quartercircle in any position.

The ends of screws are normally 45 degreechamfered.

The core diameter of the internal thread is drawn as an object line, the outer diameter as a thin unbroken line.

All lines of concealed thread are drawn as invisible edges. The thin three-quarter circle becomes a full circle shown in broken line.

2.2.4 Drawing in Three Elevations

Sometimes it is necessary to draft workpieces in three elevations to show all important parts of it (Fig. 2.16).

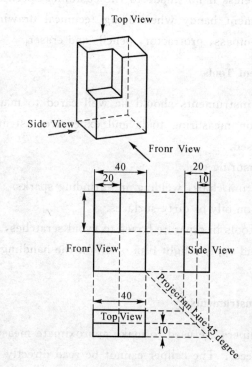

Fig. 2.16

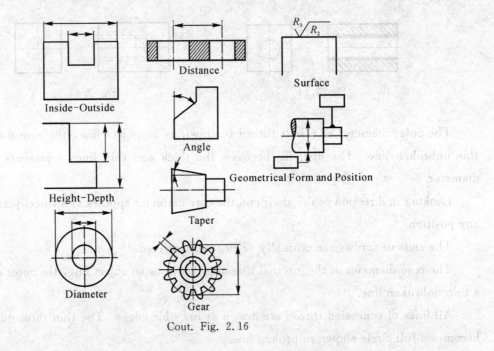

Cout. Fig. 2.16

2.3 Measuring Equipment

Industrial machining or fabrication could not go without precise measuring equipment. The parts produced are useless if not made to the customer specified dimension. You may find the following instrument handy when doing technical drawing: drawing board, T-square, drawing paper, compass, protractor, pencils and eraser.

2.3.1 Care of Measurement Tools

Measuring tools and instruments should be well cared to maintain the accuracy and quality, especially precision measuring tools and expensive instruments, otherwise their accuracy can be compromised.

(1) Never drop a measuring tool.

(2) Keepthem away from chips, welding and grinding sparks.

(3) Never placethem on oily or dirty surfaces.

(4) Store measuring tools in separate boxes to avoid scratches, nicks, or dents.

(5) Clean the tools and apply a light film of oil on the handling surfaces before putting them away.

2.3.2 Indirect Reading Instruments

Inside and outside calipers are used to make approximate measurements of the outside diameter of round workpieces. The caliper cannot be read directly and its setting must be checked with a rule or a vernier caliper (Fig. 2.17 – 2.27).

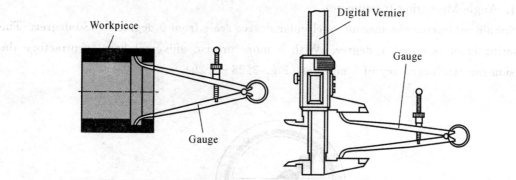

Fig. 2.17 Inside Callper with Curved Legs, a Spring, and an Adjusting Nut

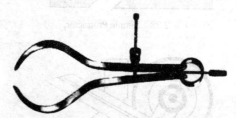

Fig. 2.18 Outside Callper with Curved Legs, a Spring, and an Adjusting Nut

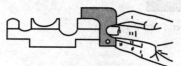

Fig. 2.19 Outside Radius Gauge

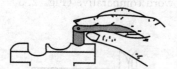

Fig. 2.20 Inside Radius Gauge

Fig. 2.21 Angle Form Gauge

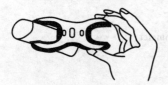

Fig. 2.22 Limit Snap Gauge

Fig. 2.23 Thread Gauge

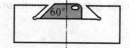

Fig. 2.24 Angle Form Gauge

Fig. 2.25 Limit Plug Gauge

Fig. 2.26 Outside Thread-Ring Gauge

Fig. 2.27 Inside Thread-Plug Gauge

— 13 —

1. Angle Measuring Instruments

Simple protractor can measure a circular degree scale from 0 degree to 180 degree. The measuring error is about 1 degree. With a more precise universal bevel protractor, the precision can reach accuracy of 5 minutes (Fig. 2.28 - 2.29).

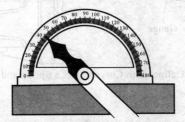

Fig. 2.28　Simple Protractor

Fig. 2.29　Universal Beverl Protractor

2. Comparative Length-Measuring Instruments

These instruments compare dimensions, hence the word comparative (Fig. 2.30 - 2.33).

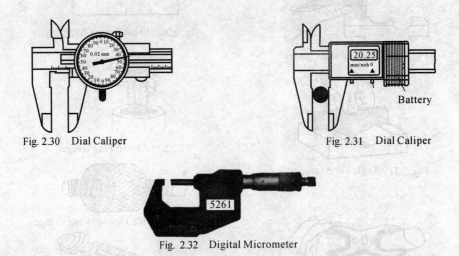

Fig. 2.30　Dial Caliper　　　　　　　　Fig. 2.31　Dial Caliper

Fig. 2.32　Digital Micrometer

Chapter 2 Technical Drawings

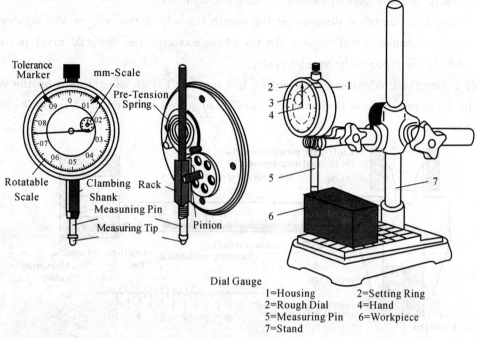

Fig. 2.33 Dial Gauge

1=Housing   2=Setting Ring
2=Rough Dial   4=Hand
5=Measuring Pin 6=Workpiece
7=Stand

2.3.3 Direct Reading Instruments

1. Steel Rules

Steel rules are often used in the metric or inch system for linear measuring. Metric rules are graduated in both millimeters and half-millimeters. Some rules are available with both inch and millimeter graduation (Fig. 2.34).

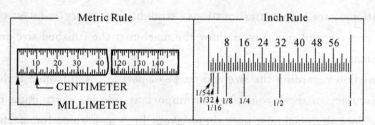

Fig. 2.34

2. Vernier Calipers

Vernier calipers are precise instruments for making internal, external and depth measurements. Both systems of metric and inch are available, and some kinds of vernier caliper provide metric readings on one side and inch readings on the other side.

Verniers for machine shops are usually 200 mm, 250 mm and 300 mm. The precision depends on the vernier scale. Common types provide an accuracy of either 0.05 mm or 0.02 mm. The example below shows an accuracy of 0.05 mm. The following Fig. 2.35

— 15 —

illustrates how to use a vernier caliper (accuracy 0.05 mm).

(1) The last numbered division on the bar to the left of the zero on the vernier scale represents the number of millimeters. In the above example the #2 (20 mm) is the last number left of the zero on the vernier scale.

(2) Count the graduations between the last number (#2) and the zero on the vernier scale. In the example above there are 8 (8 mm) graduations between the #2 and the zero on the vernier scale.

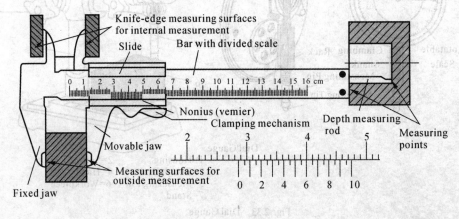

Fig. 2.35

(3) Locate the line on the vernier scale that aligns with a bar line. Divide the number below the line by 10. In the example above it is the line with #7 (7/10=0.7 mm).

(4) The measurement in the above example is 20 mm + 8 mm + 0.7 mm = 28.7 mm.

2.3.4 Laying Out

The operation of scribing center locations, straight lines, arcs, circles, or contour lines on the surface of a piece of metal is to show the machinist the finished size and shape of the part to be manufactured.

The information regarding the size and shape of part is taken from a technical drawing. The care and accuracy of the layout plays an important role in determining the accuracy of finished parts, since the machinist uses these layout lines as a guide for machining.

1. Layout Tools and Accessories (Table 2.2)

Table 2.2

Tools & Accessories	Details
Surface plate or marking table	A plate or a table made of cast iron or of granite. It must be adjusted absolute horizontally. Its surface must be perfect plane to ensure accurate scribing. To bring workpieces to the correct position on the marking table there are some other devices like prism, angle plate, V-blocks and parallels

(Continued)

Tools & Accessories	Details
Steel rule	Steel rules are the most common linear measuring tools and are available in the metric or inch system. Metric rules are graduated in both millimeters and half-millimeters. Some rules are available with both inch and millimeter graduation
Scriber	A scriber is a layout tool used for drawing layout lines on a workpiece. They are made of tool steel with hardened and tempered points. It is important that the point of the scriber be as sharp as possible to produce clear, thin, layout lines
Center punch	Normally ground to an angle of 90 degree. Before drilling a hole the center must be punched. To make a line more visible for cutting or oxy-acetylene cutting it is helpful to punch the line
Solid square or try-square	Is used for laying out workpiece in combination with steel rule and scriber. It is also used to check the angles and the surfaces for flatness
Divider	The divider is used to transfer length or circles to the workpiece. Dividers are available with and without fixing devices
Protractor	A simple protractor has a measuring range from 0 to 180 degree. The measuring error is around 1 degree
Surface gauge or vernier height gauge	Is normally used in combination with a surface plate and an angle plate to mark parallel lines. Using the simple type, the height can be adjusted with a steel rule
Angle plate	An angle plate is a precision L-shaped tool usually made of hardened steel. All its surfaces are ground to an accurate 90-degree angle and are square and parallel. It is used to support workpieces on a 90-degree angle during the layout process
V-blocks or prism	It is an accurate fabricated layout device to hold cylindrical workpieces during the layout process. They have one or more accurate 90-degree V-slots.

2. Layout Procedure

Laying out with Try-Square and Steel Rule.

(1) Remove all burrs from the workpiece and clean it properly.

(2) Start the layout from a square machined (or filed) surface.

(3) Use a try-square and a steel rule.

(4) Place the point of the scriber on the workpiece against the Try-square edge. Hold the scriber 15 degree inclined away from the workpiece and in the direction in which it is to be drawn (Fig. 2.36).

Simplified Metal Works

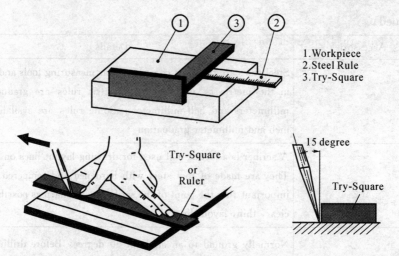

Fig. 2.36

3. Laying out Circles with the Divider (Fig. 2.37)

(1) Lay out the center of the circle.

(2) Punch the center of the circle.

(3) Adjust the divider to the proper radius while using a steel rule or a vernier caliber.

(4) Place one point of the divider in the center punch hole and give some force to this leg.

(5) Move around the fixed leg and scratch the surface.

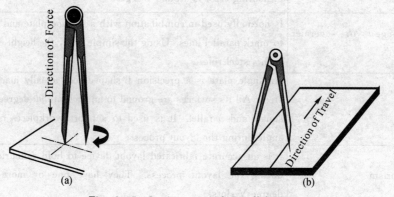

Fig. 2.37 Laying out with Parallel Lines

4. Laying out with Protractor (Fig. 2.38)

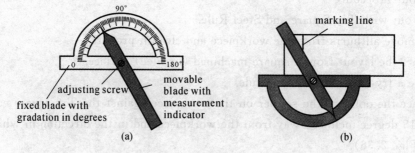

Fig. 2.38

— 18 —

Chapter 2 Technical Drawings

5. Center Punch Procedure (Fig. 2.39)

(1) Make sure that the point of the punch is sharp before starting.

(2) Hold the punch at a 45 degree angle and place the point carefully on the layout line.

(3) Tilt the punch to a vertical position and strike it gently with a light hammer.

(4) If the punch mark is not in the proper position, correct it as necessary.

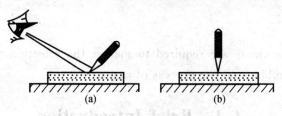

Fig. 2.39

Exercises

1. What are the common measuring tools? Can you name a few? And what are their uses?

2. What is the measurement accuracy of a vernier caliper? Can you describe how to use it?

Chapter 3 Casting

Teaching Objectives

In this chapter, students are required to master the selection of molding sands, the making of patterns, molds, and the process of casting.

3.1 Brief Introduction

Sand casting, also known as sand molded casting, is a metal casting process characterized by using sand as the mold material.

It is relatively cheap and sufficiently refractory even for steel foundry use. A suitable bonding agent (usually clay) is mixed or occurs with the sand. The mixture is moistened with water to develop strength and plasticity of the clay and to make the aggregate suitable for molding. The term "sand casting" can also refer to a casting produced via the sand casting process. Sand castings are produced in specialized factories called foundries.

Over 70% of all metal castings are produced via a sand casting process.

3.2 Basic Concepts

The basic steps involved in making sand castings are:

(1) Patternmaking. Patterns are required to make molds. The mold is made by packing molding sand around the pattern. The mold is usually made in two parts so that the pattern can be withdrawn.

In horizontal molding, the top half is called the cope, and the bottom half is called the drag.

In vertical molding, the leading half of the mold is called the swing, and the back half is called the ram.

When the patterns withdrawn from the molding material (sand or other), the imprint of the pattern provides the cavity when the mold parts are brought together. The moldcavity, together with any internal cores as required, is ultimately filled with molten metal to form the casting.

(2) If the casting is to be hollow, additional patterns, referred to as core boxes, are needed to shape the sand forms, or cores, that are placed in the mold cavity to form the interior surfaces and sometimes the external surfaces as well of the casting. Thus the void between the mold and core eventually becomes the casting.

(3) Molding is the operation necessary to prepare a mold for receiving the metal. It consists of ramming sand around the pattern placed in support, or flask, removing the pattern, setting cores in place, and creating the gating/feeding system to direct the metal into the mold cavity created by the pattern, either by cutting it into the mold by hand or by including it on thepattern, which is most commonly used.

(4) Melting and pouring are the processes of preparing molten metal of the proper composition and temperature and pouring this into the mold from transfer ladles.

(5) Cleaning includes all the operations required to remove the gates and risers that constitute the gating/feeding system and to remove the adhering sand, scale, parting fins, and other foreign material that must be removed before the casting is ready for shipment or other processing.

(6) Inspection follows, to check for defects in the casting as well as to ensure that the casting has the dimensions specified on the drawing and/or specifications. Inspection for internal defects may be quite involved, depending on the quality specified for the casting. The inspected and accepted casting sometimes is used as is, but often it is subject to further processing which may include heat treatment, painting, rust preventive oils, other surface treatment (e.g. hot-dip galvanizing), and machining. Final operations may include electrodeposited plated metals for either cosmetic oroperational requirements.

3.3 Pattern

The pattern is the principal tool during the casting process. It is the replica of the object to be made by the casting process, with some modifications. The main modifications are the addition of pattern allowances, and the provision of core prints. If the casting is to be hollow, additional patterns called cores are used to create these cavities in the finished product. The quality of the casting produced depends upon the material of the pattern, its design, and construction. The costs of the pattern and the related equipment are reflected in the cost of the casting. The use of an expensive pattern is justified when the quantity of castings required is substantial.

3.3.1 Functions of the Pattern

(1) A pattern prepares a mold cavity for the purpose of making a casting.

(2) A pattern may contain projections known as core prints if the casting requires a core and need to be made hollow.

(3) Runner, gates, and risers used for feeding molten metal in the mold cavity may form a part of the pattern.

(4) Patterns properly made and having finished and smooth surfaces reduce casting defects.

(5) A properly constructed pattern minimizes the overall cost of the castings.

3.3.2 Pattern Allowances

Pattern allowance is a vital feature as it affects the dimensional characteristics of the casting. Thus, when the pattern is produced, certain allowances must be given on the sizes specified in the finished component drawing so that a casting with the particular specification can bemade. The selection of correct allowances greatly helps to reduce machining costs and avoid rejections. The allowances usually considered on patterns and core boxes are as follows:

(1) Shrinkage or contraction allowance.
(2) Draft or taper allowance.
(3) Machining or finish allowance.
(4) Distortion or camber allowance.
(5) Rapping allowance.

1. Shrinkage or Contraction Allowance

Almost all cast metals shrink or contract volumetrically on cooling. The metal shrinkage is of two types:

(1) Liquid Shrinkage: it refers to the reduction in volume when the metal changes from liquid state to solid state at the solidus temperature. To account for this shrinkage; riser, which feed the liquid metal to the casting, are provided in the mold.

(2) Solid Shrinkage: it refers to the reduction in volume caused when metal loses temperature in solid state. To account for this, shrinkage allowance is provided on the patterns. The rate of contraction with temperature is relative to the material. For example, steel contracts to a higher degree compared to aluminum. To compensate the solid shrinkage, a shrink rule must be used in laying out the measurements for the pattern. A shrink rule forcast iron is 1/8 inch longer per foot than a standard rule. If a gear blank of 4 inches in diameter was planned to produce out of cast iron, the shrink rule in measuring it 4 inches would actually measure $4 + 1/24$ inches, thus compensating for the shrinkage (Table 3.1).

Table 3.1

Material	Dimension	Shrinkage allowance inch
Grey Cast Iron	Up to 2 feet	0.125
	2 feet to 4 feet	0.105
	over 4 feet	0.083
Cast Steel	Up to 2 feet	0.251
	2 feet to 6 feet	0.191
	over 6 feet	0.155
Aluminum	Up to 4 feet	0.155
	4 feet to 6 feet	0.143
	over 6 feet	0.125
Magnesium	Up to 4 feet	0.173
	Over 4 feet	0.155

2. Draft or Taper Allowance

By draft is meant the taper provided by the pattern maker on all vertical surfaces of the pattern so that it can be removed from the sand without tearing away the sides of the sandmold.

The Figure below shows a pattern having no draft allowance being removed from the pattern. In this case, till the pattern is completely lifted out, its sides will remain in contact with the walls of the mold, thus tending to break it (Fig. 3.1).

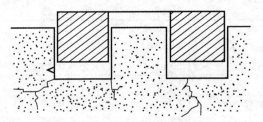

Fig. 3.1

The photo below is an illustration of a pattern having proper draft allowance. Here, the moment the pattern lifting commences, all of its surfaces are well away from the sand surface. Thus the pattern can be removed without damaging the mold cavity (Fig. 3.2).

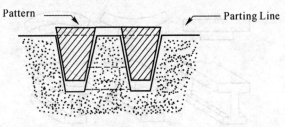

Fig. 3.2

Draft allowance varies with the complexity of the sand job. But in general inner details of the pattern require higher draft than outer surfaces. The amount of draft depends upon the length of the vertical side of the pattern to be extracted; the intricacy of the pattern; the method ofmolding; and pattern material.

3. Machining or Finish Allowance

The finish and accuracy achieved in sand casting are generally poor. When the casting is functionally required to be of good surface finish or dimensionally accurate, it is generally achieved by subsequent machining. The amount of machining allowance is affected by the method of molding and casting used viz. hand molding or machine molding, sand castingor metal mold casting. The amount of machining allowance is also affected by the size and shape of the casting; the casting orientation; the metal; and the degree of accuracy and finish required. The machining allowances recommended for different metal is given in Table below.

4. Distortion or Camber Allowance

Sometimes castings get distorted during solidification, due to their typical shape. For example, if the casting has the form of the letter U, V, T, or L etc. it will tend to contract at the closed end causing the vertical legs to look slightly inclined. This can be prevented by making the legs of the U, V, T, or L shaped pattern converge slightly (inward) so that the casting after distortion will have its sides vertical (Table 3.2).

Table 3.2

Metal	Dimension(inch)	Allowance(inch)
Cast iron	Up to 12	0.12
	12 to 20	0.20
	20 to 40	0.25
Cast steel	Up to 6	0.12
	6 to 20	0.25
	20 to 40	0.30
Non ferrous	Up to 8	0.09
	8 to 12	0.12
	12 to 40	0.16

The distortion in casting may occur due to internal stresses. These internal stresses are caused on account of unequal cooling of different section of the casting and hindered contraction (Fig. 3.3).

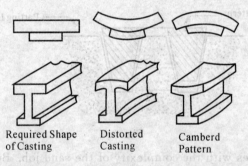

Fig. 3.3

Measure taken to prevent the distortion in casting includes:
(1) Modification of casting design.
(2) Providing sufficient machining allowance to cover the distortion effect.
(3) Providing suitable allowance on the pattern, called camber or distortion allowance (inverse reflection).

5. Rapping Allowance (Fig. 3.4)

Before the withdrawal from the sand mold, the pattern is rapped all around the vertical faces to enlarge the mold cavity slightly, which facilitate its removal. Since it enlarges

Fig. 3.4

the finalcasting made, it is desirable that the original pattern dimension should be reduced to account for this increase. There is no sure way of quantifying this allowance, since it is highly dependent on the foundry personnel practice involved.

3.3.3 Types of Pattern

Patterns are of various types, each satisfying certain casting requirements.

A pattern for a part can be made many different ways, which are classified into the following main types:

Fig. 3.5

1. Solid Pattern (Fig. 3.5)

A solid pattern is a model of the part as a single piece. It is the easiest to fabricate, but can cause some difficulties in making the mold. The parting line and runner system must be determined separately. Solid patterns are typically used for geometrically simple parts that are produced in low quantities.

2. Split Pattern (Fig. 3.6)

split pattern models the part as two separate pieces that meet along the parting line of the mold. Using two separate pieces allows the mold cavities in the cope and drag to be made separately and the parting line is already determined. Split patterns are typically used for parts that are geometrically complex and are produced inmoderate quantities.

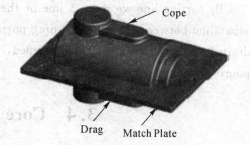

Fig. 3.6

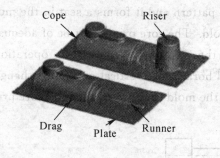

Fig. 3.7

3. Match-plate Pattern (Fig. 3.7)

A match-plate pattern is similar to a split pattern, except that each half of the pattern is attached to opposite sides of a single plate. The plate is usually made from wood or metal. This pattern design ensures proper alignment of the mold cavities in the cope and drag and the runner system can be included on the match plate. Match-plate patterns are used for larger production quantities and are often used when the process is automated.

4. Cope and Drag Pattern

A cope and drag pattern is similar to a match plate pattern, except that each half of the pattern is attached to a separate plate and the mold halves are made independently. Just as with a match plate pattern, the plates ensure proper alignment of the mold cavities in the cope and drag and the runner system can be included on the plates. Cope and drag patterns are often desirable for larger castings, where a match-plate pattern would be too heavy and cumbersome. They are also used for larger productionquantities and are often used when the process is automated.

5. Gated Patterns (Fig. 3.8)

A gated pattern is a loose pattern that has the gating systemincluded as a part of the pattern. This eliminates the time andinconsistency associated with hand-cutting the gates and runners. Also, since the gating system is designed and fabricated as part ofthe pattern, the consistency of molten metal flow into the castingand feeding of the casting during solidification is improved. Gatedpatterns are appropriate for pouring small quantities of castings when quick turn around and low cost are important.

Fig. 3.8

6. Parting Line

By parting line we mean a line or the plane of a pattern corresponding to the point of separation between the cope and drag portions of a sand mold. Parting lines must be flat or drafted so that the mold can be opened, the pattern can be removed and then closed for pouring without damage to the sand.

3.4 Core and Core Prints

Castings are often required to have holes, recesses, etc. of various sizes and shapes. The seimpressions can be obtained by using cores. So where coring is required, provision should be made to support the core inside the mold cavity. Core prints are used to serve this purpose. The core print is an added projection on the pattern and it forms a seat in the mold on which the sand core rests during pouring of the mold. The core print must be of adequate size and shape so that it can support the weight of the core during the casting operation. Depending upon the requirement a core can be placed horizontal, vertical and can be hanged inside the mold cavity. A typical job, its pattern and the mold cavity with core and core print is shown in the following figure (Fig. 3.9).

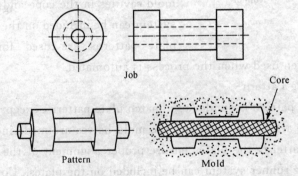

Fig. 3.9 A Typical Job, its Pattern and the Mold Cavity

3.4.1 Core and Core Box

A core is a preformed baked sand or green sand aggregate inserted in a mold to shape the interior part of a casting which cannot be shaped by the pattern.

A core box is a wood or metal structure, the cavity of which has the shape of the desired core which is made therein.

A core box, like a pattern is made by the pattern maker.

Cores run from extremely simple to extremely complicated.

A core could be a simple round cylinder form needed to core a hole through a hub of a wheel or it could be a very complicated core used to core out the water cooling channels in a cast iron engine block along with the inside of the cylinders.

Dry sand cores are for the most part made of sharp, clay-free, dry silica sand mixed with a binder and baked until cured; the binder cements the sand together (Fig. 3.10).

When the metal is poured the core holds together long enough for the metal to solidify, then the binder is finely cooked, from the heat of the casting, until its bonding power is lost or burned out. If the core mix is correct for the job, it can be readily removed from the castings interior by simply pouring it out as burnt core sand. This characteristic of a core mix is called its collapsibility.

The size and pouring temperature of a casting determines how well and how long the core will stay together.

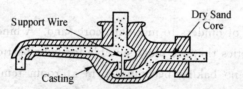

Fig. 3.10 Dry Sand Core with Support Wire

The gases generated within the core during pouring must be vented to the outside of the mold preventing gas porosity and a defect known as a core blow.

Also, a core must have sufficient hot strength to be handled and used properly.

The hot strength refers to its strength while being heated by the casting operation. Because of the shape and size of some cores they must be further strengthened with rods and wires (Fig. 3.11).

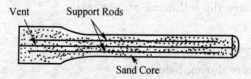

Fig. 3.11

A long span core for a length of cast iron pipe would require rodding to prevent the core from sagging or bending upward when the mold is poured because of the liquid metal exerting a strong pressure during pouring.

3.4.2 Manufacturing of Core

Core sand mixes can be done in a Muller or paddle type mixer and in small amounts on the bench by hand. The core is made by ramming the sand into the core box and placing the core on a core plate to bake (Fig. 3.12).

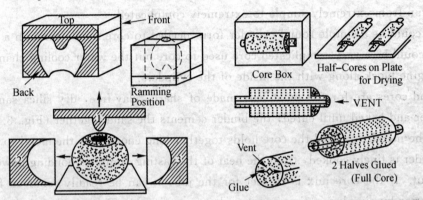

Fig. 3.12　Typical Examples of the Making of Three-part Core and Pasted Core

3.5　Binders

There are many types of binders to mix with core sand. A binder should be selected on the basis of the characteristics that are most suitable for your particular use.

Some binders require no baking becoming firm at room temperature such as rubber cement, Portland cement and sodium silicate or water glass.

In large foundry operations and in some small foundries, sodium silicate is a popular binder as it can be hardened almost instantly by blowing carbon dioxide gas through the mixture.

Oil binders require heating or baking before they develop sufficient strength to withstand the molten metal.

Sulfite binders also require heating.

There are many liquid binders made from starches, cereals and sugars.

A good binder will have the following properties:
- Strength.
- Collapse rapidly when metal starts to shrink.
- Will not distort core during baking.
- Maintain strength during storage time.
- Absorb a minimum of moisture when in the mold or in storage.
- Withstand normal handling.

- Disperse properly and evenly throughout the sand mix.
- Should produce a mixture that can be easily formed.

3.6 Molding Process and Materials

Good castings cannot be made without good molds.

The term molding process refers to the method of making the mold and the materials used.

The term casting process conveys a broader meaning, often including the molding process, the method of introducing the metal into the mold cavity, or all processes used in making the casting.

Molding processes have certain features in common.
- The use of pattern.
- Some type of aggregate mixture comprising a granular refractory and binders.
- A means of forming the aggregate mixture around the pattern.
- Hardening of aggregate or developing its bond while in contact with the pattern.
- Withdrawal of the pattern from the hardened aggregate mold.
- Assembly of the mold and core pieces to make a complete mold, metal then being poured into the mold.

3.6.1 Classification of Molding Processes

Molding processes can be classified in a number of ways. Broadly they are classified either on the basis of the method used or on the basis of the mold material used.

1. Classification Based on the Mold Material Used
   (1) Sand molding.
   ①Green sand mold.
   ②Dry sand mold, Skin dried mold.
   ③Cement bonded sand mold.
   ④Carbon-dioxide mold.
   ⑤Shell mold.
   (2) Plaster molding.
   (3) Metallic molding.
2. Classification Based on the Method Used
   (1) Bench molding.
   (2) Floor molding.
   (3) Pit molding.
   (4) Machine molding.

3.7 Kinds of Molding Sand

Molding sands can also be classified according to their use into number of varieties which are described below.

1. Green Sand

Green sand is also known as tempered or natural sand which is a just prepared mixture of silica sand with 18%-30% clay, having moisture content from 6%-8%. The clay and water furnish the bond for green sand. It is fine, soft, light and porous. Green sand is damp, when squeezed, it retains the shape and the impression of the hands under pressure.

Molds prepared by this sand are not requiring backing and hence are known as green sand molds. This sand is easily available and it possesses low cost. It is commonly employed for production of ferrous and non-ferrous castings.

2. Dry Sand

Green sand that has been dried or baked in suitable oven after the making mold and cores, is called dry sand. It possesses more strength, rigidity and thermal stability. It is mainly suitable for larger castings. Mold prepared in this sand are known as dry sand molds.

3. Loam Sand

Loam is mixture of sand and clay with water to a thin plastic paste. Loam sand possesses high clay as much as 30%-50% and 18% water. Patterns are not used for loam molding and shape is given to mold by sweeps. This is particularly employed for loam molding used for large grey iron castings.

4. Facing Sand

Facing sand forms the face of the mold, gives surface finish to casting. It is next to the surface of the pattern and comes into contact with molten metal when the mold is poured. Initial coating will form around the pattern and hence give the molding surface. This sand is subjected to severest conditions and must possess, therefore, high strength refractoriness. It is made of silica sand and clay.

5. Backing Sand

Backing sand or floor sand is used to back up the facing sand and is used to fill the whole volume of the molding flask. Used molding sand is mainly employed for this purpose. The backing sand is sometimes called black sand because that repeatedly used molding sand is black in color due to burningwhen contact with the molten metal.

6. Parting Sand

Parting sand without binder and moisture is used to keep the green sand not to stick to the pattern and also to allow the sand on the parting surface of the cope and drag to separate without clinging.

7. Core Sand

Core sand is used for making cores and it is sometimes also known as oil sand. This is highly rich silica sand mixed with oil binders such as core oil which composed of linseed oil, resin, light mineral oil and other bind materials.

3.7.1 The basic Properties of Molding Sand and Core Sand

1. Refractoriness

Refractoriness is defined as the ability of molding sand to withstand high temperatures without breaking down or fusing thus facilitating to get sound casting. It is a highly important characteristic of molding sands. Refractoriness can only be increased to a limited extent.

Molding sand with poor refractoriness may burn on to the casting surface and no smoothcasting surface can be obtained. The degree of refractoriness depends on the SiO_2 i.e. quartz content, and the shape and grain size of the particle. The higher the SiO_2 content and the rougher the grain volumetric composition, the higher is the refractoriness of the molding sand and core sand. Refractoriness is measured by the sinter point of the sand rather than its melting point.

2. Permeability

It is also termed as porosity of the molding sand in order to allow the escape of any air, gasesor moisture present or generated in the mold when the molten metal is poured into it. All these gaseous matter generated during pouring and solidification process must escape, otherwise the casting becomes defective. Permeability is a function of grain size, grain shape, and moisture and clay contents in the molding sand. The extent of ramming of the sand directly affects the permeability of the mold. Permeability of mold can be further increased by venting using vent rods.

3. Cohesiveness

It is property of molding sand by virtue which the sand grain particles interact and attract each other within the molding sand. Thus, the binding capability of the molding sand gets enhanced to increase the green, dry and hot strength property of molding and core sand.

4. Green Strength

The green sand after water has been mixed into it, must have sufficient strength and toughness to permit the making and handling of the mold. For this, the sand grains must be adhesive, i.e. they must be capable of attaching themselves to another body, and therefore, the sand grains having high adhesiveness will cling to the sides of the molding box. Also, the sand grains must have the property known as cohesiveness, i.e. ability of the sand grains to stick to one another. By virtue of this property, the pattern can be taken out from the mold without breaking the mold and also the erosion of mold wall surfaces does not occur during the flow of molten metal. The green strength also depends upon the grain shape and size, amount and type of clay and the moisture content.

5. Dry Strength

As soon as the molten metal is poured into the mold, the moisture in the sand layer adjacent to the hot metal gets evaporated and this dry sand layer must have sufficient strength to its shape in order to avoid erosion of mold wall during the flow of molten metal.

6. Flowability or Plasticity

It is the ability of the sand to get compacted and behave like a fluid. It will flow uniformly to all portions of pattern when rammed and distribute the ramming pressure evenly all around in all directions. Generally sand particles resist moving around corners or projections. In general, flow ability increases with decrease in green strength, and, decrease in grain size. The flowability also varies with moisture and clay content.

7. Adhesiveness

It is property of molding sand to get stick or adhere with foreign material such sticking of molding sand with inner wall of molding box.

8. Collapsibility

After the molten metal in the mold gets solidified, the sand mold must be collapsible so that free contraction of the metal occurs and this would naturally avoid the tearing or cracking of the contracting metal. In absence of this property the contraction of the metal is hindered by the mold and thus results in tears and cracks in the casting. This property is highly desired in cores.

9. Miscellaneous Properties

In addition to above requirements, the molding sand should not stick to the casting and should not chemically react with the metal. Molding sand should be cheap and easily available. It should be reusable for economic reasons. Its coefficient of thermal expansion should be sufficiently low.

3.8 Classification Based on the Mold Material

Sand Molding

Molding processes where a sand aggregate is used to make the mold produce by far the largest quantity of castings. Whatever the metal poured into sand molds, the product may be called a sand casting.

(1) Green-sand molding: Among the sand-casting processes, molding is mostly done with green sand. Green molding sand may be defined as a plastic mixture of sand grains, clay, water, and other materials which can be used for molding and casting processes. The sand is called "green" because of the moisture present and is thus distinguished from dry sand. The basic steps in green-sand molding are as follows:

1) Preparation of the pattern. Most green-sand molding is done with match plate or cope and drag patterns. Loose patterns are used when relatively few castings of a type are to be made. In simple hand molding the loose pattern is placed on a mold board and surrounded with a suitable-sized flask.

2) Making the mold. Molding requires the ramming of sand around the pattern. As the

sand is packed, it develops strength and becomes rigid within the flask. Ramming may be done by hand. Both cope and drag are molded in the same way, but the cope must provide for the sprue. The gating-system parts of the mold cavity are simply channels for the entry of the molten metal.

3) Core setting. With cope and drag halves of the mold made and the pattern withdrawn, cores are set into the mold cavity to form the internal surfaces of the casting.

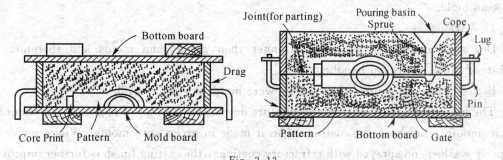

Fig. 3.13

4) Closing and weighting. With cores set, the cope and drag are closed. The cope must usually be weighted down or clamped to the drag to prevent it from floating when the metal is poured. Because of the nature of green-sand molding and molding sands, the process has certain advantages and limitations.

5) Advantages:

Great flexibility as a production process. Mechanical equipment can be utilized for performing molding and its allied operations. Furthermore, green sand can be reused many times by reconditioning it with water, clay, and ether materials. The molding process can be rapid and repetitive.

Usually, the most direct route from pattern to mold ready for pouring is by green-sand molding.

Economy. Green sand molding is ordinarily the least costly method of molding.

6) Limitations in the use of green-sand molding are:

Some casting designs require the use of other casting processes. Thin, long projections of green sand in a mold cavity are washed away by the molten metal or may not even be moldable.

Cooling fins on air-cooled-engine cylinder blocks and head are an example. Greater strength is then required of the mold.

Certain metals and some castings develop defects if poured into molds containing moisture.

The dimensional accuracy and surface finish of green-sand castings may not be adequate.

Large castings require greater mold strength and resistance to erosion than are available in green sands.

(2) Dry-sand Molds: Dry-sand molds are actually made with molding sand in the green

condition. The sand mixture is modified somewhat to favor good strength and other properties after the mold is dried. Dry-sand molding may be done the same way as green-sand molding on smaller sizes of castings. Usually, the mold-cavity surface is coated or sprayed with a mixture which, upon drying, imparts greater hardness or refractoriness to the mold. The entire mold is then dried in an oven at 300 to 650 F or by circulating heated air through the mold. The time-consuming drying operation is one inherent disadvantage of the dry-sand mold.

1) Advantages.

Dry sand molds are generally stronger than green sand molds and therefore can withstand much additional handling.

Better dimension control than if they were molded in green sand.

The improved quality of the sand mixture due to the removal of moisture can result in a much smoother finish on the castings than if made in green sand molds. Where molds are properly washed and sprayed with refractory coatings, the casting finish is further improved.

2) Disadvantages.

This type of molding is much more expensive than green sand molding and is not a high production process. Correct baking (drying) times are essential.

(3) Skin-dried Molding: The effect of a dry-sand mold may be partially obtained by drying the mold surface to some depth, 1/4 to 1 inch. Skin drying may be performed by torches or electrical heating elements directed at the mold surface. Skin-dried molds must be poured shortly after drying, so that moisture from the undried sand will not penetrate the dried skin.

3.9 Steps Involved in Making a Sand Mold

(1) First, a suitable size of molding box is selected for a two-piece pattern. Care should also be taken in such sense that the molding box must adjust mold cavity, riser and the gating system (sprue, runner and gates etc.).
(2) Next, place the drag portion of the pattern with the parting surface down on thebottom board.
(3) The facing sand is then sprinkled carefully all around the pattern so that the pattern does not stick with molding sand during withdrawn of the pattern.
(4) The drag is then filled with loose prepared molding sand and ramming of the molding sand is done uniformly in the molding box around the pattern. Fill the molding sand once again and then perform ramming. Repeat the process three four times.
(5) The excess amount of sand is then removed using strike off bar to bring molding sand at the same level of the molding flask height to complete the drag.
(6) The drag is then rolled over and the parting sand is sprinkled over on the top of the drag.
(7) Now the cope pattern is placed on the drag pattern and alignment is done using dowel pins.
(8) Then cope (flask) is placed over the rammed drag and the parting sand is sprinkled all

Chapter 3 Casting

around the cope pattern.

(9) Sprue and riser pins are placed in vertically position at suitable locations using support of molding sand. It will help to form suitable sized cavities for pouring molten metal etc.

(10) Fill the cope with molding sand and ram uniformly.

(11) Strike off the excess sand from the top of the cope.

(12) Remove sprue and riser pins and create vent holes in the cope with a vent wire.

(13) Sprinkle parting sand over the top of the cope surface and roll over the cope on the bottom board.

(14) Rap and remove both the cope and drag patterns and repair the mold suitably if needed and dressing is applied.

(15) The gate is then cut connecting the lower base of sprue basin with runner and then the mold cavity.

(16) Bake the mold in case of a dry sand mold.

(17) Set the cores in the mold, if needed and close the mold by inverting cope over drag.

(18) The cope is then clamped with drag and the mold is ready for pouring. Fig. 3.14.

Fig. 3.14

Exercises

1. What are the basic steps of casting?
2. What is pattern allowance? Can you classify different pattern allowances?
3. List different types of pattern.
4. What are the properties of molding sand?

Chapter 4 Forging

Teaching Objectives

This chapter provides fundamentals of metal working process for forging in order to understand mathematical approaches used in the calculation of applied forging loads required to cause plastic deformation to give the final product.

Classification of metal forging methods is also provided with descriptions of defects observed from the forging processes. The solutions to tackle such defects will also be addressed.

4.1 Brief Introduction

Forging is the working of metal into a useful shape by hammering or pressing. It is one of the oldest metal working arts. In ancient times, blacksmiths use hammers and anvils to make various iron weapons and farm tools. Today, forging machines are capable of making parts ranging in size of a bolt to a turbine rotor. Most forging operations are carried out hot, though certain metals may be cold-forged. Metals and alloys have the ability to be formed into useful shapes by plastic deformation. In this process they also develop a wide range of properties particularly strength and toughness. The working of the metals into shapes by means of forging methods refines grain structure, develops inherent strength giving characteristics, improves physical properties and produces structural uniformity free from hidden internal defects. Moreover, the special feature of flow lines along the contour of the forging produce marked directional properties.

4.2 Forging Processes

4.2.1 Temperature

For forging, a metal must be heated to a temperature at which it will possess high plastic properties both at the beginning and at the end of the forging process, e. g. temperature to begin the forging for soft, low carbon steels is $1,250-1,300℃$ and finishing temperature is $800-850℃$. If forging operation is finished at lower temperature, this leads to cold hardening and cracks. With excessive heating the forgings suffer oxidation and much metal is wasted.

When steel is heated well above the upper critical temperature, the grains begins to grow in size and they will continue to grow as the temperature is increased. During forging, the grains are broken up and become finer. If forging temperature is high (above 910℃) the grains will grow during process of cooling in the air, the cold forging will then have a coarse grained structure a low mechanical properties. If forging is finished at low temperature (below 910℃), the grains will not grow when the steel is cooled owing to the low temperature. The cold forging will then possess a fine grained structure and high mechanical properties.

If steel is hammered when it is below the lower critical temperature (above 723℃) it will be cold worked and may be given small hair cracks.

Table 4.1

	Forging Must	
	Start at	Finish at
Mild Steel	1,300℃	800℃
Medium Carbon Steel	1,250℃	750℃
High Carbon Steel	1,150℃	800℃
Wrought Iron	1,250℃	900℃
Aluminum & Magnesium Alloys	500℃	300℃
Copper, Brass and Bronze	950℃	600℃

4.2.2 Surface Color for Iron and Steel

Table 4.2

Color	Temperature (℃)
Faire Red	500
Blood Red	650
Cherry Red	750
Bright Red	850
Salmon	900
Orange	950
Yellow	1,050
White	1,200

4.2.3 Optimum Fire Colors When Forging Metal Workpiece

Table 4.3

Sparkling or sizzling heat	Never used. Sparkling or sizzling heat should be avoided. Bright sparks are flying off the material and the effected area is making a sizzling noise. The sizzling is literally the steel burning. At this point the steel should be considered destroyed, and the effected area is of no further use. Cut off the effected area from the bar and throw it away. To try to save it or use it will end in failure of the piece because the grain structure is destroyed and the metal will crumble during forging and be brittle when cooled. Avoid sizzling heat
White heat or welding heat	For welding mild steel and iron. Remove from fire when sparks begin to fly and stop forge welding when down to light yellow heat
Light yellow heat	Best heat for forging wrought iron and mild steel. Try to obtain this heat before removing from the fire for forging. Sometimes this is described as a light welding heat
Yellow heat	For forging wrought iron and mild steel
Bright red or light red (orange) heat	Forging of alloy steels is done at this color. Forging of mild steel and wrought iron at this temperature is beginning to become more difficult than at yellow heat
Full red heat	For final straightening of wrought iron and mild steel. Put back in fire at this heat
Dark red or dull red heat	For finish forging which means smoothing out dents and sharp corners. Do not forge alloy steels below this temperature. Mild steel and wrought iron may still be worked somewhat
Black heat or dark blood red heat	Not used for working iron. Black heat describes a temperature range from an almost undetectable red heat (not bright enough to be visible except in a darkened place) to well below red-heat. For safety reasons all iron and tools not glowing red hot, should be treated as black hot until proven otherwise. New smiths should take note that any metal that is still cooling and no longer red hot is still dangerous. And all tools which come into contact with the hot metal are also heated and dangerous

4.2.4 Heating Devices

For heating the workpiece, there are many kinds of heating furnaces to heat the workpiece to optimum temperature.

Rotary-Hearth Furnace

These are doughnut shaped and are set the rotate slowly so that the stock is heated to the correct temperature during one rotation. These are also heated by gas or oil (Fig. 4.1).

Fig. 4.1

Box or Batch Type Furnaces

These are widely employed in forging shops for heating small and medium size stock because they are the least expensive. There is a great variety of design of box-type furnace, each differing in their location of their charging doors, firing devices and methods employed for discharging their products. These furnaces are usually constructed of a rectangular steel frame, may be 2 400 mm widely 1200 mm deep, lined with insulating and refractory bricks. One or more burners for gas or oil are provided on the sides. The work-pieces are placed side-by-side inside a low 'slot' through which the furnace operator reaches with tong. This is therefore sometimes called slot-type furnace. However, usually two people tend all types of furnaces, one filling in the cold stock and other bringing heated stock to the forge operator (Fig. 4.2).

Fig. 4.2

Resistance Furnace

Resistance furnaces are faster than induction furnaces and are often automated. In resistance heating the stock is connected into the circuit of a step down transformed. Fixtures must be made for holding each different length, shape and diameter of stock. However, the fixtures are often quite simple, and some can be adjusted to handle a 'family' of parts (Fig. 4.3).

Fig. 4.3

Continuous or Conveyor Furnaces

These are used of several types if only one end of the work must be heated though they also will heat complete stock. Especially for larger stock, a pusher furnace may be used. This has an air or oil-operated cylinder to push stock end to end through a narrow furnace. The pieces are charged at one end, are conveyed through the furnace, and are moved at the other end at the correct temperature for forging (Fig. 4.4).

Fig. 4.4

Induction Furnaces

In induction furnaces the stocks are passed through induction coils in the furnaces. These furnaces are becoming very popular because induction greatly decreases scale, can often be operated by one person, requires less maintenance than oil-or gas-fired furnaces, and is faster. Delivery to the forging machine operator can be effected by slides or automatic handling equipment (Fig. 4.5).

Fig. 4.5

4.2.5 Possible Heating Defects

Heating improper defects could be caused:

(1) Decarburization. The decarburization means the metal surface of the carbon is oxidized at high temperatures, making the surface carbon content less than the internal carbon content.

(2) Overheating. The metal billets are heated over required temperature, or residence time in the forging lasts too long, which leads to coarse crystalline texture.

Overheating will lead to decrease of the impact toughness, which will lead to fissure of workpiece during forging.

(3) Overburnt. The metal billet is heated near the melting point or stay too long in the high-temperature. Oxidation of iron occurs which in turn lead to the brittleness of workpiece.

(4) Crack. The uneven distribution of temperature on the ingot leads to different stresses of the structure, and fissure happens.

(5) Oxidation. The iron content of the ingot surface gets oxidized and which in turn damages the quality of the workpiece, and reduces its service life.

4.3 Classification of Forging Operations

Forging processes can be broadly classified as the following types.

4.3.1 Cold vs. Hot Forging

Hot or warm forging — advantage: reduction in strength and increase in ductility of work metal, hot forging require smaller forces, but not has good accuracy or surface.

Cold forging — advantage: increased strength due to strain hardening.

4.3.2 Impact vs. Press Forging

Forge hammer — applies an impact force.

Forge press — applies gradual force.

4.3.3 Open Die Forging, Closed Die Forging and Flashless Forging

1. Open Die Forging

This process consists of hammering heated metal between a pair of flat dies to make metal flow laterally with minimum constraint. For obtaining the desired rough shapes the skill of the operator in manipulating the work piece is essential. At times some helping tools are used to improve the shapes and for special purposes. This is suitable to the type of production where quantities are small and sizes of work are large. The type of products produced on these open frame type pneumatic power hammers are shafts, rings, discs, large gear blanks etc (Fig. 4.6).

— Deformation operation reduces height and increases diameter of work
— Common names include upsetting or upset forging

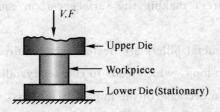

Fig. 4.6

Upsetting: a process of open-die forging, which reduces the height and increases the diameter of workpieces. Pancaking or barreling can occur due to friction forces at the die-workpiece interface or thermal differences (Fig. 4.7).

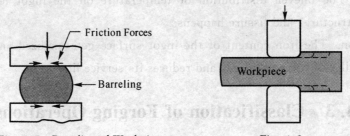

Fig. 4.7 Barreling of Workpiece   Fig. 4.8

Cogging: also called drawing out, that is successively reducing the thickness of a bar with open die forging. This process doesn't need excessive forces or machining. The right figure illustrate the cogging process. Blacksmiths use this process to reduce the thickness of bars by hammering the part on an anvil (Fig. 4.8).

2. Impression Die Forging (closed-die forging)

This consists of hammering or pressing heated bars or billets within close dies. The die contains a cavity or impression, the workpiece is deformed under high pressure in a closed cavity. These impressions impart the desired shape to the work pieces. The process provides precision forging with close dimensional tolerance. Closed dies are expensive. This process also creates flash (excess metal, which squirts out of the cavity as a thick ribbon of metal).

In view of the many variables in this process such as material temperature, die condition, machine load, complexity of job, die lubrication etc., the process consists of several stages of manufacture: billet heating, fullering, bending or flattening (if necessary), blocker or mould, and finally finish forging (Fig. 4.9).

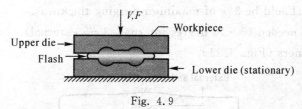

Fig. 4.9

Billet heating (to around 1250℃) is done in forge furnaces which may be either box type pusher, walking beam, rotary hearth or induction furnaces.

Reduce-rolling is done to distribute the metal to suit forging shape in a separate rolling machine. Flattening, bending, blocking and finish forging operations are done in the main forging unit i.e. Press or a Hammer. Trimming and hot coining operations are done on auxiliary units. Heat treatment, shot blasting, grinding etc. are the finishing operations, which are done in separate equipment (Fig. 4.10).

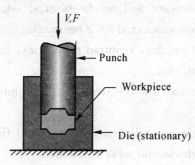

Fig. 4.10

3. Flashless Forging

The workpiece is completely constrained in die and there is no creation of excess flash.

Flashless forging uses precise volumes of material to completely fill the die cavity. Undersize workpieces will not fill the cavity, oversize workpieces will cause high pressures and may damage the dies. The advantage of this methods lies in simper operations and no wasted material.

4.4 Die Design

The designer must know workpiece material's strength and ductility, deformation rate, temperature sensitivity, frictional characteristics, and forgeability (capability to undergo deformation without cracking), he should also be able to evaluate the shape, size and complexity of his design. Material usually flows in the direction of least resistance, which is why intermediate shapes may need to be formed. An experienced designer will implement pre-shaping method to prevent material from easily flowing into flash, to produce favorable grain flow patterns, and minimize friction and wear at the die-workpiece interface. The use of computers can model and predict material flow.

Parting lines are usually at the largest cross section.

Flash clearance should be 3% of maximum forging thickness.

Draft angles are needed (7 - 10 degree internal, 3 - 5 external).

Avoid sharp corners (Fig. 4.11).

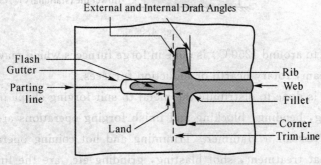

Fig. 4.11

Die materials must have strength and toughness at elevated temperatures, hardenability, resistance to mechanical or thermal shock, and wear resistance.

Die selection can be based on size, required properties, forging temperature, operation type, cost and production quantities.

Common die materials are tool and die steels with chromium, nickel, molybdenum and vanadium.

Dies are forged from castings, and then machined and finished as needed, often with heat treatment to increase hardness and wear resistance.

Lubricants act as thermal barriers for hot workpiece and cooler dies, improve metal flow, and are parting agents.

Improper design, the use of defective material, faulty heat treatment and finishing operations, overheating can cause cracks, excessive wear, overloading, misuse or improper handling all lead to die failures, which are often catastrophic at high speeds and therefore, dangerous to employees.

4.5 Type of Forging Machines

4.5.1 The Following Illustrates Various Presses

(1) Hydraulic press.

(2) Mechanical press with an eccentric drive; the eccentric shaft can be replaced by

a crankshaft
  (3) Knuckle-joint press.
  (4) Screw press.
  (5) Gravity drop hammer.
Fig. 4.12.

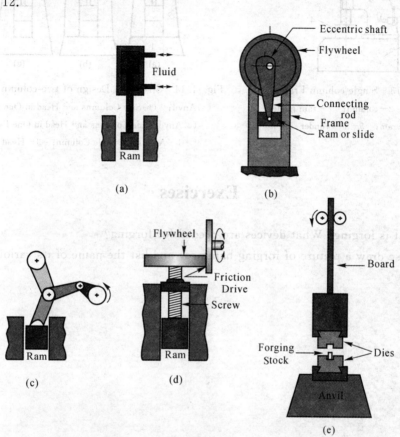

Fig. 4.12

4.5.2 Forging Hammers

(1) Gravity Drop Hammers Drop forging with a free falling ram; energy based on product of ram's weight and drop height.

(2) Power Drop Hammers The downstroke ramming is accelerated by steam, air, or hydraulic pressure.

(3) Counterblow Hammers Two rams that approach each other horizontally or vertically (part may be rotated between successive blows); operate at high speeds with less vibrations transmitted — very large capacity possible.

(4) High Energy Rate Machines Ram accelerated by an inert gas at high pressure; very high speeds but problems with maintenance, die breakage and safety.

Fig. 4.13 and Fig. 4.14.

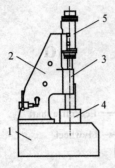
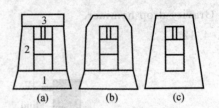

Fig. 4.13 Single-column Frame
1—anvil; 2—column; 3—ram guide;
4—ram; 5—air cylinder

Fig. 4.14 Structural Design of two-column Frames
(a) Anvil; (b) Side Columns and Head in One Piece;
(c) Anvil, Side Columns and Head in One Piece
1—Anvil; 2—Side Column; 3—Head

Exercises

1. What is forging? What devices are needed for forging?
2. Please draw a picture of forging hammer, and list the name of its various parts.

Chapter 5   Introduction to Welding

Teaching Objectives

After this chapter, students should be able to get familiar with the classification of welding, basic welding operation procedures and use of welding techniques.

5.1   Brief Introduction

The term welding refers to the process of joining metals by heating them to their melting temperature and causing the molten metal to flow together. Welding is widely used in the fabrication, maintenance, and repair of parts and structures in industry. While there are many methods for joining metals, welding is one of the most convenient and rapid methods available. These range from simple steel brackets to nuclear reactors.

Welding, like any skilled trade, is broad in scope. A skill need practice and experience as well as patience. This chapter is designed to equip the learner with a background of basic information applicable to welding in general.

5.2   Welding Process

The earliest known form of welding, called forge welding, dates back to the year 2000 B.C. Forge welding is a primitive process of joining metals by heating and hammering until the metals are fused (mixed) together. Although forge welding still exists, it is mainly limited to the blacksmith trade.

Today, there are many welding processes available. Fig. 5.1 provides a list of processes used in modern metal fabrication and repair. This list, published by the American Welding Society (AWS), shows the official abbreviations for each process. For example, RSW stands for resistance spot welding. Shielded metal arc welding (SMAW) is an arc-welding process that fuses (melts) metal by heating it with an electric arc created between a covered metal electrode and the metals being joined. Of the welding processes listed in Fig. 5.1, shielded metal arc welding, called stick welding, is the most common welding process. The primary differences between the various welding processes are the methods by which heat is generated to melt the metal.

Once you understand the theory of welding, you can apply it to most welding processes. The most common types of welding are oxyfuel gas welding (OFW), arc welding (AW), and resistance welding (RW). As a beginner, your primary concern is gas and arc welding.

The primary difference between these two processes is the method used to generate the heat.

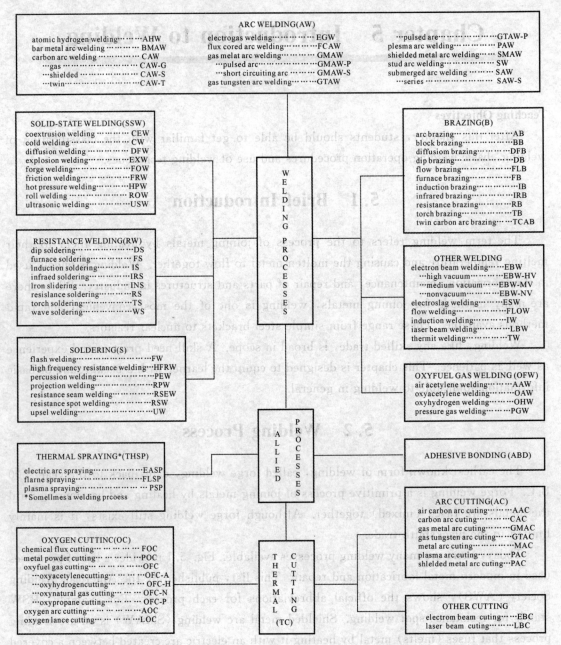

Fig. 5.1 Welding Processes

5.2.1 Gas Welding

One of the most popular welding methods uses a gas flame as a source of heat. In the oxyfuel gas welding process (Fig. 5.2), heat is produced by burning a combustible gas, such as MAPP or acetylene, mixed with oxygen. Gas welding is widely used in maintenance and repair work because of the ease in transporting oxygen and fuel cylinders. The oxyfuel process is adaptable to brazing, cutting, and heat treating all types of metals.

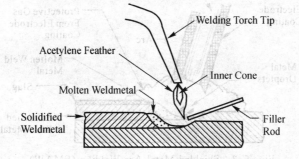

Fig. 5.2 Oxyfuel Gas Welding (OFW)

5.2.2 Arc Welding

Arc welding uses an electric arc to join the metals being welded. A distinct advantage of arc welding over gas welding is the concentration of heat. In gas welding the flame spreads over a large area, sometimes causing heat distortion. The concentration of heat, characteristic of arc welding, is an advantage because less heat spread reduces buckling and warping.

This heat concentration also increases the depth of penetration and speeds up the welding operation; therefore, is often more practical and economical than gas welding.

All arc-welding processes have three things in common: a heat source, filler metal, and shielding. The source of heat in arc welding is produced by the arcing of an electrical current between two contacts. The power source is called a welding machine or simply, a welder. This should not be confined with the same term that is also used to describe the person who is performing the welding operation. The welder (welding machine) is either electric or motor-powered.

5.2.3 Shielded Metal Arc Welding (SMAW)

Shielded metal arc welding (Fig. 5.3) is performed by striking an arc between a coated-metal electrode and the base metal. Once the arc has been established, the molten metal from the tip of the electrode flows together with the molten metal from the edges of the base metal to form a sound joint. This process is known as fusion.

The coating from the electrode forms a covering over the weld deposit, shielding it from contamination; therefore the process is called shielded metal arc welding. The main advantages of shielded metal arc welding are that high-quality welds are made rapidly at a low cost.

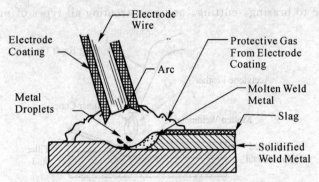

Fig. 5.3 Shielded Metal Arc Welding (SMAW)

5.2.4 Gas Shielded Arc Welding

The primary difference between shielded metal arc welding and gas shielded arc welding is the type of shielding used. In gas shielded arc welding, both the arc and the molten puddle are covered by a shield of inert gas. The shield of inert gas prevents atmospheric contamination, thereby producing a better weld. The primary gases used for this process are helium, argon, or carbon dioxide. In some instances, a mixture of these gases is used. The processes used in gas shielded arc welding are known as gas tungsten arc welding (GTAW) (Fig. 5.4) and gas metal arc welding (GMAW) (Fig. 5.5). You will also hear these called "TIG" and "MIG." Gas shielded arc welding is extremely useful because it can be used to weld all types of ferrous and nonferrous metals of all thicknesses.

Now that we have discussed a few of the welding processes available, which one should you choose?

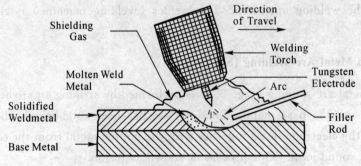

Fig. 5.4 Gas Tungsten Arc Welding (GTAW)

There are no hard-and-fast rules. In general, the controlling factors are the types of metal you are joining, cost, nature of the products you are fabricating, and the techniques you use to fabricate them. Because of its flexibility and mobility, gas welding is widely used for maintenance and repair work in the field. On the other hand, you should probably choose gas shielded metal arc welding to repair a critical piece of equipment made from aluminum or stainless steel.

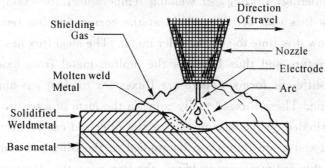

Fig. 5.5 Gas Metal Arc Welding (GMAW)

5.3 Filler Metals

When welding two pieces of metal together, there is often a space between the joint. The material that you add to fill this space during the welding process is known as the filler metal. Two types of filler metals commonly used in welding are welding rods and welding electrodes. Welding rod refers to a form of filler metal that does not conduct an electric current during the welding process. The only purpose of a welding rod is to supply filler metal to the joint. This type of filler metal is often used for gas welding. In electric-arc welding, the term electrode refers to the component that conducts the current from the electrode holder to the metal being welded. Electrodes are classified into two groups: consumable and nonconsumable. Consumable electrodes not only provide a path for the current but also supply filler metal to the joint. An example is the electrode used in shielded metal-arc welding. Nonconsumable electrodes are only used as a conductor for the electrical current, such as in gas tungsten arc welding. The filler metal for gas tungsten arc welding is a hand fed consumable welding rod.

5.4 Fluxes

Before performing any welding process, you must ensure the base metal is clean. No matter how much the base metal is physically cleaned, it still contains impurities. These impurities, called oxides, result from oxygen combining with the metal and other contaminants in the base metal. Unless these oxides are removed by using a proper flux, a faulty weld may result. The term flux refers to a material used to dissolve oxides and release trapped gases and slag (impurities) from the base metal; thus the flux can be thought of as a cleaning agent. In performing this function, the flux allows the filler metal and the base metal to be fused.

Different types of fluxes are used with different types of metals; therefore, you should choose a flux formulated for a specific base metal. Beyond that, you can select a flux based on the expected soldering, brazing, or welding temperature; for example, when brazing, you should select a flux that becomes liquid at the correct brazing temperature. When it melts, you will know it is time to add the filler metal. The ideal flux has the right fluidity at the welding temperature and thus blankets the molten metal from oxidation. Fluxes are available in many different forms. There are fluxes for oxyfuel gas applications, such as brazing and soldering. These fluxes usually come in the form of a paste, powder, or liquid. Powders can be sprinkled on the base metal, or the filler rod can be heated and dipped into the powder. Liquid and paste fluxes can be applied to the filler rod and to the base metal with a brush. For shielded metal arc welding, the flux is on the electrode. In this case, the flux combines with impurities in the base metal, floating them away in the form of a heavy slag which shields the weld from the atmosphere.

No single flux is satisfactory for universal use; however, there are a lot of good general-purpose fluxes for use with common metals. In general, a good flux has the following characteristics:

1) It is fluid and active at the melting point of the fuller metal.

2) It remains stable and does not change to a vapor rapidly within the temperature range of the welding procedure.

3) It dissolves all oxides and removes them from the joint surfaces.

4) It adheres to the metal surfaces while they are being heated and does not ball up or blow away.

5) It does not cause a glare that makes it difficult to see the progress of welding or brazing.

6) It is easy to remove after the joint is welded.

7) It is available in an easily applied form.

Caution

Nearly all fluxes give off fumes that may be toxic. Use ONLY in well-ventilated spaces. It is also good to remember that ALL welding operations require adequate ventilation whether a flux is used or not.

5.5　Weld Joints

The weld joint is where two or more metal parts are joined by welding. The five basic types of weld joints are the butt, corner, tee, lap, and edge, as shown in Fig. 5.6.

A butt joint is used to join two members aligned in the same plane (Fig. 5.6, view A). This joint is frequently used in plate, sheet metal, and pipe work. A joint of this type may be either square or grooved. Some of the variations of this joint are discussed later in this chapter.

Corner and tee joints are used to join two members located at right angles to each other

(Fig. 5.6, views B and C). In cross section, the corner joint forms an L-shape, and the tee joint has the shape of the letter T.

Various joint designs of both types have uses in many types of metal structures.

Alap joint, is made by lapping one piece of metal over another (Fig. 5.6, view D). This is one of the strongest types of joints available; however, for maximum joint efficiency, you should overlap the metals a minimum of three times the thickness of the thinnest member you are joining. Lap joints are commonly used with torch brazing and spot welding applications.

An edge joint is used to join the edges of two or more members lying in the same plane. In most cases, one of the members is flanged, as shown in figure 5.6, view E. While this type of joint has some applications in platework, it is more frequently used in sheet metal work. An edge joint should only be used for joining metals 1/4 inch or less in thickness that are not subjected to heavy loads.

The above paragraphs discussed only the five basic types of joints; however, there are many possible variations. Later in this chapter, we discuss some of these variations.

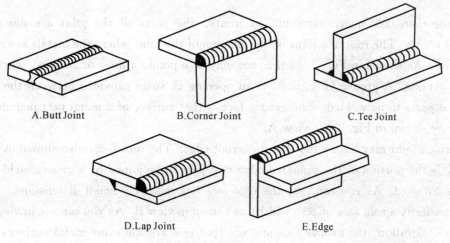

Fig. 5.6   Basic Weld-joints

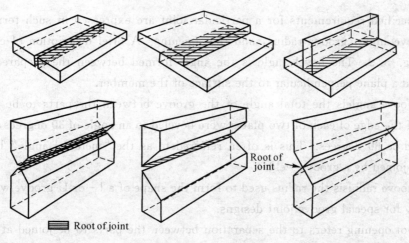

Fig. 5.7   Root of Joints

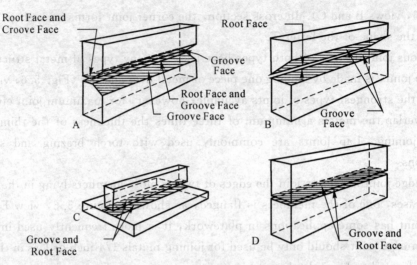

Fig. 5.8 The Groove Face, Root Face, and Root Edge of Joints

Parts of Joints

While there are many variations of joints, the parts of the joint are described by standard terms. The root of a joint is that portion of the joint where the metals are closest to each other. As shown in Fig. 5.7, the root may be a point, a line, or an area, when viewed in cross section. A groove (Fig. 5.8) is an opening or space provided between the edges of the metal parts to be welded. The groove face is that surface of a metal part included in the groove, as shown in Fig. 5.8, view A.

A given joint may have a root face or a root edge. The root face, also shown in Fig. 5.8 view A, is the portion of the prepared edge of a part to be joined by a groove weld that has not been grooved. As you can see, the root face has relatively small dimensions. The root edge is basically a root face of zero width, as shown in view B. As you can see in views C and D of the illustration, the groove face and the root face are the same metal surfaces in some joints.

The specified requirements for a particular joint are expressed in such terms as bevel angle, groove angle, groove radius, and root opening. A brief description of each term is shown inFig. 5.9. The bevel angle is the angle formed between the prepared edge of a member and a plane perpendicular to the surface of the member.

The groove angleis the total angle of the groove between the parts to be joined. For example, if the edge of each of two plates were beveled to an angle of 30 degrees, the groove angle would be 60 degrees. This is often referred to as the "included angle" between the parts to be joined by a groove weld.

The groove radiusis the radius used to form the shape of a J - or U-groove weld joint. It is used only for special groove joint designs.

The root opening refers to the separation between the parts to be joined at the root of the joint. It is sometimes called the "root gap".

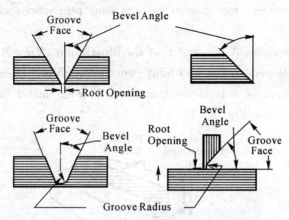

Fig. 5.9 Bevel Angle, Groove Angle, Groove Radius and Root Opening of Joints for Welding

To determine the bevel angle, groove angle, and root opening for a joint, you must consider the thickness of the weld material, the type of joint to be made, and the welding process to be used. As a general rule, gas welding requires a larger groove angle than manual metal-arc welding. The root opening is usually governed by the diameter of the filler material. This, in turn, depends on the thickness of the base metal and the welding position.

Having an adequate root opening is essential for root penetration.

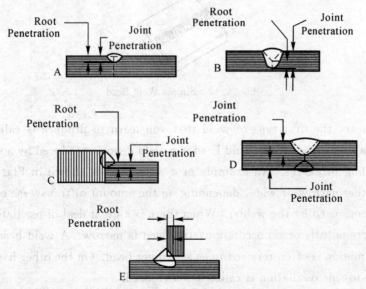

Fig. 5.10 Root Penetration and Joint Penetration of Welds

Root penetration and joint penetration of welds are shown in Fig. 5.10. Root penetration refers to the depth that a weld extends into the root of the joint. Root penetration is measured on the center line of the root cross section. Joint penetration refers to the minimum depth that a groove (or a flange) weld extends from its face into a joint, exclusive of weld reinforcement. As you can see in the figure, the terms, root penetration and joint penetration, often refer to the same dimension.

This is the case in views A, C, and E of the illustration. View B, however, shows the difference between root penetration and joint penetration. View D shows joint penetration only. Weld reinforcement is a term used to describe weld metal in excess of the metal necessary to fill a joint (See Fig. 5.11).

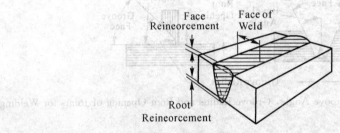

Fig. 5.11 Weld Reinforcement

5.6 Types of Welds

There are many types of welds. Some of the common types you will work with are the bead, groove, fillet, surfacing, tack, plug, slot and resistance.

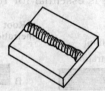

Fig. 5.12 Simple Weld Bead

As a beginner, the first type of weld that you learn to produce is called a weld bead (referred to simply as a bead). A weld bead is a weld deposit produced by a single pass with one of the welding processes. An example of a weld bead is shown in Fig. 5.12. A weld bead may be either narrow or wide, depending on the amount of transverse oscillation (side-to-side movement) used by the welder. When there is a great deal of oscillation, the bead is wide; when there is little or no oscillation, the bead is narrow. A weld bead made without much weaving motion is often referred to as a stringer bead. On the other hand, a weld bead made with side-to-side oscillation is called a weave bead.

Groove welds are simply welds made in the groove between two members to be joined. The weld is adaptable to a variety of butt joints, as shown in Fig. 5.13. Groove welds may be joined with one or more weld beads, depending on the thickness of the metal. If two or more beads are deposited in the groove, the weld is made with multiple-pass layers, as shown in Fig. 5.14. As a rule, a multiple-pass layer is made with stringer beads in manual operations. Groove welds are frequently used.

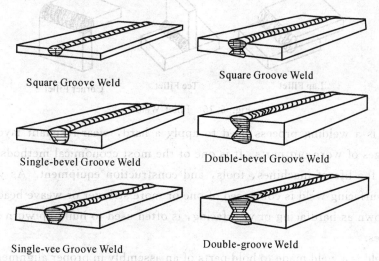

Fig. 5.13 Standard Groove Welds

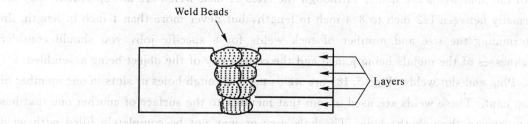

Fig. 5.14 Multiple-pass Layers

As shown in Fig. 5.15. Buildup sequence refers to the order in which the beads of a multiple-pass weld are deposited in the joint.

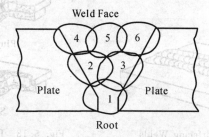

Fig. 5.15 Weld Layer Sequence

NOTE: Often welding instructions specify an interpass temperature. The interpass temperature refers to the temperature below which the previously deposited weld metal must be before the next pass may be started.

Later in the chapter, you will understand the significance of the buildup sequence and the importance of controlling the interpass temperature.

Across-sectional view of a fillet weld (Fig. 5.16) is triangular in shape. This weld is used to join two surfaces that are at approximately right angles to each other in a lap, tee or corner joint.

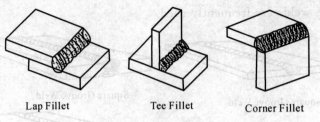

Lap Fillet   Tee Fillet   Corner Fillet

Fig. 5.16 Fillet Welds

Surfacing is a welding process used to apply a hard, wear-resistant layer of metal to surfaces or edges of worn-out parts. It is one of the most economical methods of conserving and extending the life of machines, tools, and construction equipment. As you can see in Fig. 5.17, a surfacing weld is composed of one or more stringer or weave beads. Surfacing, sometimes known as hardfacing or wearfacing, is often used to build up worn shafts, gears, or cutting edges.

A tack weld is a weld made to hold parts of an assembly in proper alignment temporarily until the final welds are made. Although the sizes of tack welds are not specified, they are normally between 1/2 inch to 3/4 inch in length, but never more than 1 inch in length. In determining the size and number of tack welds for a specific job, you should consider thicknesses of the metals being joined and the complexity of the object being assembled.

Plug and slot welds (Fig. 5.18) are welds made through holes or slots in one member of a lap joint. These welds are used to join that member to the surface of another one that has been exposed through the hole. The hole may or may not be completely filled with weld metal. These types of welds are often used to join face-hardened plates from the backer soft side, to install liner metals inside tanks, or to fill up holes in a plate.

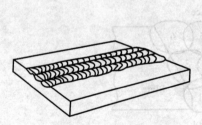

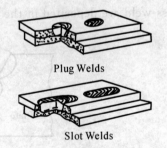

Plug Welds

Slot Welds

Fig. 5.17 Surfacing Welds   Fig. 5.18 Plug and Slot Welds

Resistance welding is a metal fabricating process in which the fusing temperature is generated at the joint by the resistance to the flow of an electrical current. This is accomplished by clamping two or more sheets of metal between copper electrodes and then passing an electrical current through them. When the metals are heated to a melting temperature, forging pressure is applied to weld the pieces together. Spot and seam welding (Fig. 5.19) are two common types of such processes.

Spot welding is probably the most commonly used type of resistance welding. The material to be joined is placed between two electrodes and pressure is applied. Next, a charge of electricity is sent from one electrode through the material to the other electrode. Spot welding is especially useful in fabricating sheet metal parts.

Seam weldingis like spot welding except that the spots overlap each other, making a continuous weld seam. In this process, the metal pieces pass between roller types of electrodes. As the electrodes revolve, the current is automatically turned on and off at the speed at which the parts are set to move. This type of welding is most often used in industrial manufacturing.

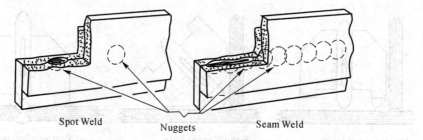

Fig. 5.19 Spot and Seam Welds

Fig. 5.20 shows a groove weld and a fillet weld. The face is the exposed surface of a weld on the side from which the weld was made. The toe is the junction between the face of the weld and the base metal. The root of a weld includes the points at which the back of the weld intersects the base metal surfaces. When we look at a triangular cross section of a fillet weld, as shown in view B, the leg is the portion of the weld from the toe to the root. The throat is the distance from the root to a point on the face of the weld along a line perpendicular to the face of the weld. Theoretically, the face forms a straight line between the toes.

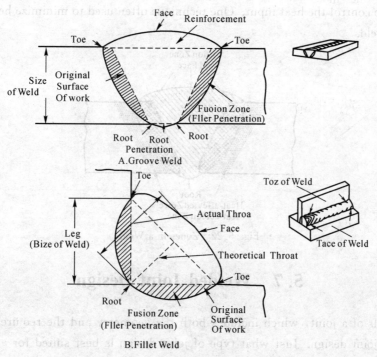

Fig. 5.20 Parts of a Groove Weld and Fillet Weld

NOTE: The terms leg and throat apply only to fillet welds.

In determining the size of a groove weld (Fig. 5.20, view A), such factors as the depth of the groove, root opening, and groove angle must be taken into consideration. The size of a fillet weld (view B) refers to the length of the legs of the weld. The two legs are assumed to be equal in size unless otherwise specified.

A gauge used for determining the size of a weld is known as a welding micrometer. Fig. 5.21 shows how the welding micrometer is used to determine the various dimensions of a weld.

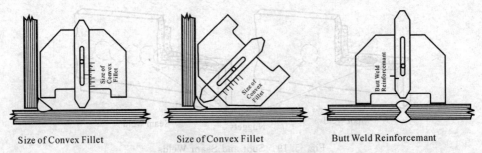

Fig. 5.21 Using a Welding Micrometer

The fusion zone, as shown in Fig. 5.22, is the region of the base metal that is actually melted. The depth of fusion is the distance that fusion extends into the base metal or previous welding pass.

Another zone of interest to the welder is the heat affected zone, as shown in Fig. 5.22. This zone includes that portion of the base metal that has not been melted; however, the structural or mechanical properties of the metal have been altered by the welding heat.

Because the mechanical properties of the base metal are affected by the welding heat, it is important to control the heat input. One technique often used to minimize heat input is the intermittent weld.

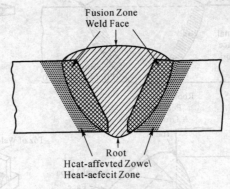

Fig. 5.22 Zones in a Weld

5.7 Welded Joint Design

The details of a joint, which includes both the geometry and the required dimensions, are called the joint design. Just what type of joint design is best suited for a particular job depends on many factors. Although welded joints are designed primarily to meet strength and safety requirements, there are other factors that must be considered. A few of these factors are as follows:

Whether the load will be in tension or compression and whether bending, fatigue, or impact stresses will be applied.

How a load will be applied; that is, whether the load will be steady, sudden, or variable.

The direction of the load as applied to the joint.

The cost of preparing the joint.

Another consideration that must be made is the ratio of the strength of the joint compared to the strength of the base metal. This ratio is called joint efficiency. An efficient joint is one that is just as strong as the base metal.

Earlier in this chapter, we discussed the five basic types of welded joints — butt, corner, tee, lap, and edge. While there are many variations, every joint you weld will be one of these basic types. Now, we will consider some of the variations of the welded joint designs and the efficiency of the joints.

Butt Joints

The square butt joint is used primarily for metals that are 3/16 inch or less in thickness. The joint is reasonably strong, but its use is not recommended when the metals are subject to fatigue or impact loads. Preparation of the joint is simple, since it only requires matching the edges of the plates together; however, as with any other joint, it is important that it is fitted together correctly for the entire length of the joint. It is also important that you allow enough root opening for the joint. Fig. 5.23 shows an example of this type of joint.

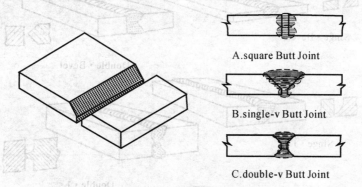

Fig. 5.23 Butt Joints

When you are welding metals greater than 3/16 inch in thickness, it is often necessary to use agrooved butt joint. The purpose of grooving is to give the joint the required strength. When you are using a grooved joint, it is important that the groove angle is sufficient to allow the electrode into the joint; otherwise, the weld will lack penetration and may crack.

However, you also should avoid excess beveling because this wastes both weld metal and time. Depending on the thickness of the base metal, the joint is either single-grooved (grooved on one side only) or double-grooved (grooved on both sides). As a welder, you primarily use the single-V and double-V grooved joints.

The single-V butt joint (Fig. 5.23, view B) is for use on plates 1/4 inch through 3/4 inch in thickness. Each member should be beveled so the included angle for the joint is approximately 60 degrees for plate and 75 degrees for pipe. Preparation of the joint requires a special beveling machine (or cutting torch), which makes it more costly than a square butt joint. It also requires more filler material than the square joint; however, the joint is stronger than the square butt joint. But, as with the square joint, it is not recommended when subjected to bending at the root of the weld.

The double-V butt joint (Fig. 5.23, view C) is an excellent joint for all load conditions. Its primary use is on metals thicker than 3/4 inch but can be used on thinner plate where strength is critical. Compared to the single-V joint, preparation time is greater, but you use less filler metal because of the narrower included angle. Because of the heat produced by welding, you should alternate weld deposits, welding first on one side and then on the other side. This practice produces a more symmetrical weld and minimizes warpage.

Remember, to produce good quality welds using the groove joint, you should ensure the fit-up is consistent for the entire length of the joint, use the correct groove angle, use the correct root opening, and use the correct root face for the joint. When you follow these principles, you produce better welds every time. Other standard grooved butt joint designs include the bevel groove, J-groove, and U-groove, as shown in Fig. 5.24.

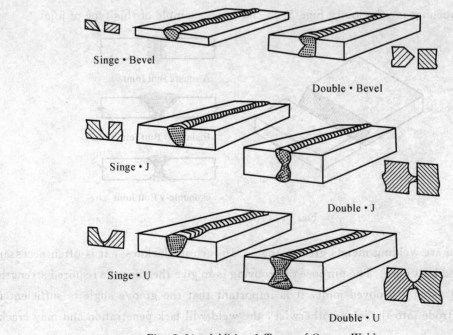

Fig. 5.24 Additional Types of Groove Welds

Corner Joints

The flush corner joint (Fig. 5.25, view A) is designed primarily for welding sheet metal that is 12 gauge or thinner. It is restricted to lighter materials, because deep penetration is sometimes difficult and the design can support only moderate loads.

The half-open corner joint (Fig. 5.25, view B) is used for welding materials heavier than 12 gauge. Penetration is better than in the flush corner joint, but its use is only recommended for moderate loads. The full-open corner joint (Fig. 5.25, view C) produces a strong joint, especially when welded on both sides. It is useful for welding plates of all thicknesses.

Tee Joints

The square tee joint (Fig. 5.26, view A) requires a fillet weld that can be made on one or both sides. It can be used for light or fairly thick materials. For maximum strength, considerable weld metal should be placed on each side of the vertical plate.

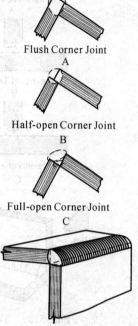

Fig. 5.25   Corner Joints

The single-bevel tee joint(Fig. 5.26, view B) can withstand more severe loadings than the square tee joint, because of better distribution of stresses. It is generally used on plates of 1/2 inch or less in thickness and where welding can only be done from one side.

The double-bevel tee joint (Fig. 5.26, view C) is for use where heavy loads are applied and the welding can be done on both sides of the vertical plate.

Lap Joints

The single-fillet lap joint (Fig. 5.27, view A) is easy to weld, since the filler metal is simply deposited along the seam. The strength of the weld depends on the size of the fillet. Metal up to 1/2 inch in thickness and not subject to heavy loads can be welded using this joint. When the joint will be subjected to heavy loads, you should use the double-fillet lap joint (Fig. 5.27, view B). When welded properly, the strength of this joint is very close to the strength of the base metal.

Simplified Metal Works

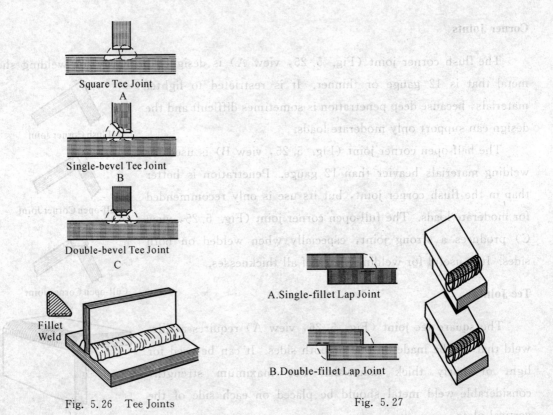

Fig. 5.26 Tee Joints

Fig. 5.27

Edge Joints

The flanged edge joint (Fig. 5.28, view A) is suitable for plate 1/4 inch or less in thickness and can only sustain light loads. Edge preparation for this joint may be done, as shown in either views B or C.

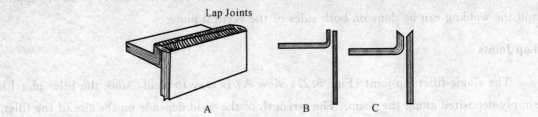

Fig. 5.28

5.8 Welding Positions

All welding is done in one of four positions: ①flat, ②horizontal, ③vertical, or ④overhead. Fillet or groove welds can be made in all of these positions. Fig. 5.29 shows the various position used in plate welding. The American Welding Society (AWS) identifies

Chapter 5　Introduction to Welding

these positions by a number designation; for instance, the 1G position refers to a groove weld that is to be made in the flat position here the 1 is used to indicate the flat position and the G indicates a groove weld. For a fillet weld made in the flat position, the number/letter designation is 1F (F for fillet). These number/letter designations refer to test positions. These are positions a welder would be required to use during a welding qualification test. As a Steelworker, there is a good possibility that someday you will be required to certify or perform a welding qualification test; therefore, it is important that you have a good understanding and can apply the techniques for welding in each of the test positions.

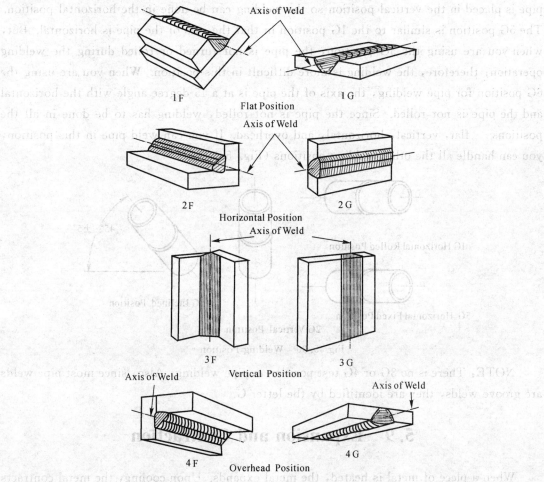

Fig. 5.29　Welding Positions

Because of gravity, the position in which you are welding affects the flow of molten filler metal. Use the flat position, if at all possible, because gravity draws the molten metal downward into the joint making the welding faster and easier. Horizontal welding is a little more difficult, because the molten metal tends to sag or flow downhill onto the lower plate. Vertical welding is done in a vertical line, usually from bottom to top; however, on thin material downhill or downhand welding may be easier. The overhead position is the most difficult position. Because the weld metal flows downward, this position requires considerable practice on your part to produce good quality welds.

Although the terms flat, horizontal, vertical, and overhead sufficiently describe the positions for plate welding, they do not adequately describe pipe welding positions. In pipe welding, there are four basic test positions used (Fig. 5.30). Notice that the position refers to the position of the pipe, not the position of welding. Test position 1G is made with the pipe in the horizontal position. In this position, the pipe is rolled so that the welding is done in the flat position with the pipe rotating under the arc. This position is the most advantageous of all the pipe welding positions. When you are welding in the 2G position, the pipe is placed in the vertical position so the welding can be done in the horizontal position. The 5G position is similar to the 1G position in that the axis of the pipe is horizontal. But, when you are using the 5G position, the pipe is not turned or rolled during the welding operation; therefore, the welding is more difficult in this position. When you are using the 6G position for pipe welding, the axis of the pipe is at a 45-degree angle with the horizontal and the pipe is not rolled. Since the pipe is not rolled, welding has to be done in all the positions — flat, vertical, horizontal, and overhead. If you can weld pipe in this position, you can handle all the other welding positions (Fig. 5.30).

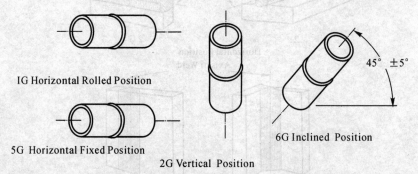

Fig. 5.30 Welding Position

NOTE: There is no 3G or 4G test position in pipe welding. Also, since most pipe welds are groove welds, they are identified by the letter G.

5.9 Expansion and Contraction

When a piece of metal is heated, the metal expands. Upon cooling, the metal contracts and tries to resume its original shape. The effects of this expansion and contraction are shown in Fig. 5.31. View A shows a bar that is not restricted in any way. When the bar is heated, it is free to expand in all directions. If the bar is allowed to cool without restraint, it contracts to its original dimensions. When the bar is clamped in a vise (view B) and heated, expansion is limited to the unrestricted sides of the bar. As the bar begins to cool, it still contracts uniformly in all directions. As a result, the bar is now deformed. It has become narrower and thicker, as shown in view C.

Chapter 5 Introduction to Welding

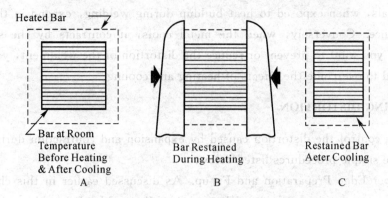

Fig. 5.31 The Effects of Expansion and Contraction

These same expansion and contraction forces act on the weld metal and base metal of a welded joint; however, when two pieces of metal are welded together, expansion and contraction may not be uniform throughout all parts of the metal. This is due to the difference in the temperature from the actual weld joint out to the edges of the joint. This difference in temperature leads to internal stresses, distortion, and warpage. Fig. 5.32 shows some of the most common difficulties that you are likely to encounter. When you are welding a single-V butt joint (Fig. 5.32, view A), the highest temperature is at the surface of the molten puddle. The temperature decreases as you move toward the root of the weld and away from the weld. Because of the high temperature of the molten metal, this is where expansion and contraction are greatest. When the weld begins to cool, the surface of the weld joint contracts (or shrinks) the most, thus causing warpage or distortion. View B shows how the same principles apply to a tee joint. Views C and D show the distortions caused by welding a bead on one side of a plate and welding two plates together without proper tack welds.

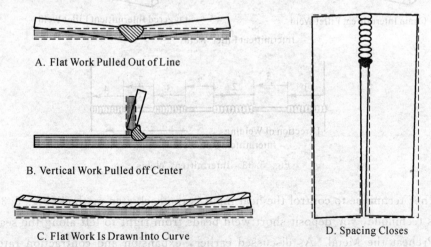

Fig. 5.32 Distortions Caused by Welding

All metals, when exposed to heat buildup during welding, expand in the direction of least resistance. Conversely, when the metal cools, it contracts by the same amount; therefore, if you want to prevent or reduce the distortion of the weldment, you have to use some method to overcome the effects of heating and cooling.

CONTROLLING DISTORTION

You can control the distortion caused by expansion and contraction during welding by following the simple procedures listed below.

(1) Proper Edge Preparation and Fit-up. As discussed earlier in this chapter, proper edge preparation and fit-up are essential to good quality welds. By making certain the edges are properly beveled and spacing is adequate, you can restrict the effects of distortion. Additionally, you should use tack welds, especially on long joints. Tack welds should be spaced at least 12 inches apart and run approximately twice as long as the thickness of the weld.

(2) Control the Heat Input. The faster a weld is made, the less heat is absorbed by the base metal. It is often necessary to use a welding technique designed to control heat input. An intermittent weld (sometimes called a skip weld) is often used instead of one continuous weld. When you are using an intermittent weld, a short weld is made at the beginning of the joint. Next, you skip to the center of the seam and weld a few inches. Then, you weld at the other end of the joint. Finally, you return to the end of the first weld and repeat the cycle until the weld is finished. Fig. 5.33 shows the intermittent welds.

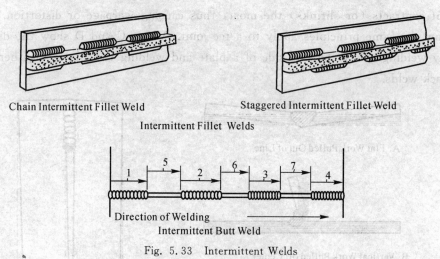

Fig. 5.33 Intermittent Welds

Another technique to control the heat input is the back-step method (Fig. 5.34). When using this technique, you deposit short weld beads from right to left along the seam.

(3) Preheat the Metal. As discussed earlier, expansion and contraction rates are not uniform in a structure during welding due to the differences in temperature throughout the metal. To control the forces of expansion and contraction, you preheat the entire structure

before welding. After the welding is complete, you allow the structure to cool slowly.

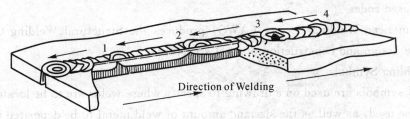

Fig. 5.34 Back-step Welding

(4) Limit the Number of Weld Passes. You can keep distortion to a minimum by using as few weld passes as possible. You should limit the number of weld passes to the number necessary to meet the requirements of the job (See Fig. 5.35).

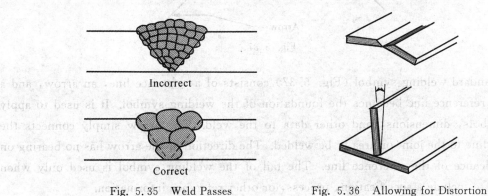

Fig. 5.35 Weld Passes   Fig. 5.36 Allowing for Distortion

(5) Use Jigs and Fixtures. Since holding the metal in a fixed position prevents excessive movements, the use of jigs and fixtures can help prevent distortion. A jig or fixture is simply a device used to hold the metal rigidly in position during the welding operation.

(5) Allow for Distortion. A simple remedy for the distortion caused by expansion and contraction is to allow for it during fit-up. To reduce distortion, you angle the parts to be welded slightly in the opposite direction in which the contraction takes place. When the metal cools, contraction forces pull the pieces back into position. Figure 5.36 shows how distortion can be overcome in both the butt and tee joints.

5.10 Welding Procedures

There are many factors involved in the preparation of any welded joint. The detailed methods and practices used to prepare a particular weldment are called the welding procedure. A welding procedure identifies all the welding variables pertinent to a particular job or project. Generally, these variables include the welding process, type of base metal, joint design, welding position, type of shielding, preheating and post-heating requirements, welding machine setting, and testing requirements.

Welding procedures are used to produce welds that will meet the requirements of commonly used codes.

The American Welding Society (AWS) produces the Structural Welding Code that is used for the design and construction of steel structures.

1. Welding Symbols

Special symbols are used on a drawing to specify where welds are to be located, the type of joint to be used, as well as the size and amount of weld metal to be deposited in the joint. These symbols have been standardized by the American Welding Society (AWS).

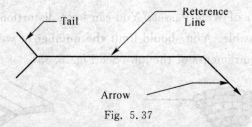

Fig. 5.37

A standard welding symbol (Fig. 5.37) consists of a reference line, an arrow, and a tail. The reference line becomes the foundation of the welding symbol. It is used to apply weld symbols, dimensions, and other data to the weld. The arrow simply connects the reference line to the joint or area to be welded. The direction of the arrow has no bearing on the significance of the reference line. The tail of the welding symbol is used only when necessary to include a specification, process, or other reference information.

2. Weld Symbols

The term weld symbol refers to the symbol for a specific type of weld. As discussed earlier, fillet, groove, surfacing, plug, and slot are all types of welds. Basic weld symbols are shown in Fig. 5.38. The weld symbol is only part of the information required in the welding symbol. The term welding symbol refers to the total symbol, which includes all information needed to specify the weld(s) required.

Basic Weld Symbols									
Bead	Fillet	Plug or Slot	Groove Or Butt						
			Square	V	Bevel	U	J	Flare V	Flare Bevel

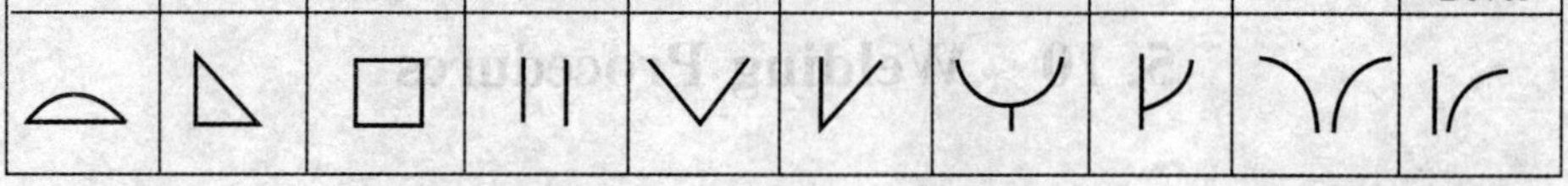

Fig. 5.38 Basic Weld Symbols

Fig. 5.39 shows how a weld symbol is applied to the reference line. Notice that the vertical leg of the weld symbol is shown drawn to the left of the slanted leg.

Regardless of whether the symbol is for a fillet, bevel, J - groove, or flare-bevel weld, the vertical leg is always drawn to the left.

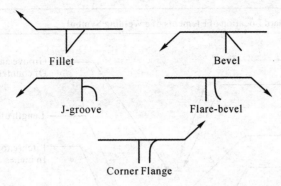

Fig. 5.39 Weld Symbols Drawn on a Reference Line

Fig. 5.40 shows the significance of the positions of the weld symbols position on the reference line. In view A the weld symbol is on the lower side of the reference line that is termed the arrow side. View B shows a weld symbol on the upper side of the reference line that is termed the other side. When weld symbols are placed on both sides of the reference line, welds must be made on both sides of the joint (view C).

When only one edge of a joint is to be beveled, it is necessary to show which member is to be beveled. When such a joint is specified, the arrow of the welding symbol points with a definite break toward the member to be beveled. This is shown in Fig. 5.41. Fig. 5.42 shows other elements that may be added to a welding symbol. The information applied to the reference line on a welding symbol is read from left to right regardless of the direction of the arrow.

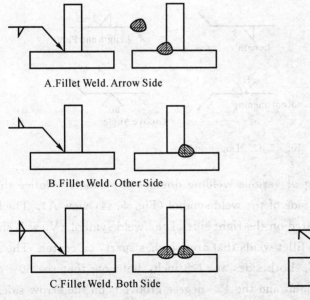

Fig. 5.40 Weld Locations

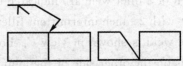

Fig. 5.41 Arrowhead Indicate Beveled Plate

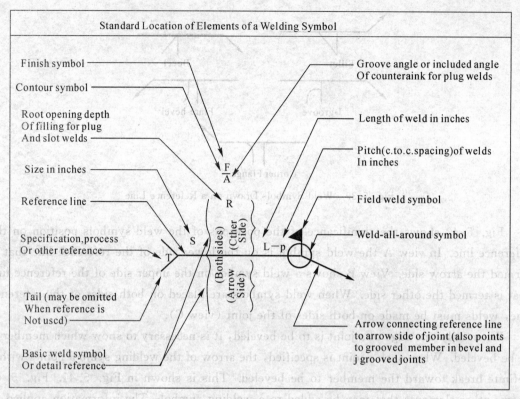

Fig. 5.42   Elements of a Welding Symbol

3. Dimensioning

Fig. 5.43 shows how dimensions are applied to symbols.

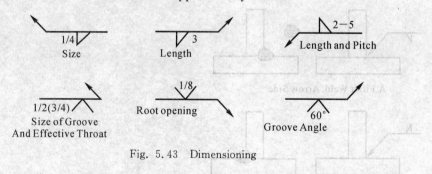

Fig. 5.43   Dimensioning

Fig. 5.44 shows the meaning of various welding dimension symbols. Notice that the size of a weld is shown on the left side of the weld symbol (Fig. 5.44, view A). The length and pitch of a fillet weld are indicated on the right side of the weld symbol. View B shows a tee joint with 2 - inch intermittent fillet welds that are 5 inches apart, on center. The size of a groove weld is shown in view C. Both sides are 1/2 inch, but note that the 60 - degree groove is on the other side of the joint and the 45 - degree groove is on the arrow side.

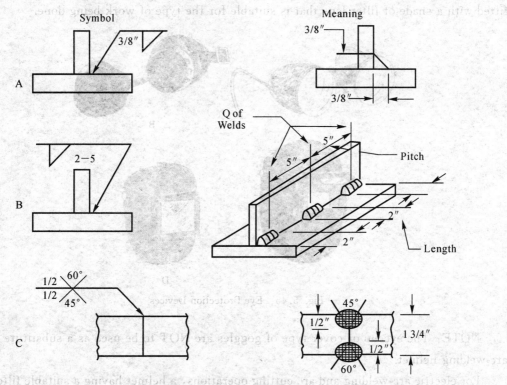

Fig. 5.44 Dimensioning

5.11 Safety

Mishaps frequently occur in welding operations. Sometimes, they result in serious injury to the welder or other personnel working in the immediate area. Mostly, mishaps occur because of carelessness, lack of knowledge, and the misuse of available equipment. In this section we'll focus on such topics as protective clothing, eye protection devices, and practices applicable to the personal safety of the operator and personnel working nearby.

Proper eye protection is of the utmost importance. Eye protection is necessary because of the hazards posed by stray flashes, reflected glare, flying sparks, and globules of molten metal. Devices used for eye protection include helmets and goggles.

NOTE: In addition to providing eye protection, helmets also provide a shield against flying metal and ultraviolet rays for the entire face and neck. Fig. 5.45 shows several types of eye protection devices in common use.

Flash goggles (view A) are worn under the welder's helmet and by persons working around the area where welding operations are taking place. This spectacle type of goggles has side shields and may have either an adjustable or nonadjustable nose bridge.

Eyecup or cover types of goggles (view B) are for use in fuel-gas welding or cutting operations. They are contoured to fit the configuration of the face. These goggles must be fitted with a shade of filter lens that is suitable for the type of work being done.

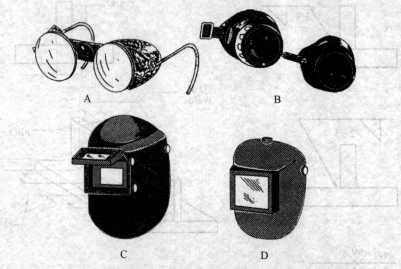

Fig. 5.45   Eye Protection Devices

NOTE: The eyecup or cover type of goggles are NOT to be used as a substitute for an arc-welding helmet.

For electric arc-welding and arc-cutting operations, a helmet having a suitable filter lens is necessary. The helmet shown in view C has an opening, called a window, for a flip-up filter lens 2 inches by 4 1/4 inches in size. The helmet shown in view D has a 4 1/2-inch by 5 1/4-inch window. The larger window affords the welder a wider view and is especially useful when the welder is working in a confined place where head and body movement is restricted. When welding in locations where other welders are working, the welder should wear flash goggles beneath his helmet to provide protection from the flashes caused by the other welders' arcs. The flash goggles will also serve as eye protection when chipping the slag from a previous weld deposit.

Helmets and welding goggles used for eye protection are made from a nonflammable insulating material. They are fitted with a removable protective colored filter and a clear cover lens.

NOTE: The purpose of the clear cover lens is to protect the filter lens against pitting caused by sparks and hot metal spatter. The clear lens must be placed on the outside of the filter lens. The clear lens should be replaced when it impairs vision.

Filter lenses are furnished in a variety of shades, which are designated by number. The lower the number, the lighter the shade; the higher the number, the darker the shade. Table 5.1 shows you the recommended filter lens shade for various welding operations. The filter lens shade number selected depends on the type of work and somewhat on the preference of the user. Remember, a filter lens serves two purposes. The first is to diminish the intensity of the visible light to a point where there is no glare and the welding area can be

Chapter 5 Introduction to Welding

clearly seen. The second is to eliminate the harmful infrared and ultraviolet radiations coming from the arc or flame; consequently, the filter lens shade number you select must not vary more than two shades from the numbers recommended in table 5.1.

Table 5.1 Recommended Filter Lenses

Shade No.	Operation
Up to 4	Light electric spot welding or for protection from stray light from nearby welding.
5	Light gas cutting and welding.
6 – 7	Gas cutting, mdium gas welding, and arc welding up to 30 amperes.
8 – 9	Heary gas welding and arc welding and cutting, 30 – 75 amperes.
10 – 11	Arc welding and cutting, 76 – 200 amperes.
12	Arc welding and cutting, 201 – 400 amperes.
13 – 14	Arc welding and cutting exceeding 400 amperes.

Rule of thumb: When selecting the proper shade of filter lens for electric-arc welding helmets, place the lens in the helmet and look through the lens as if you were welding. Look at an exposed bare light bulb and see if you can distinguish its outline. If you can, then use the next darker shade lens. Repeat the test again. When you no longer see the outline of the bulb, then the lens is of the proper shade. Remember that this test should be performed in the same lighting conditions as the welding operation is to be performed. Welding in a shop may require a shade lighter lens than if the same job were being performed in bright daylight. For field operations, this test may be performed by looking at a bright reflective object.

WARNING

Looking at the arc with the naked eye could lead to permanent eye damage. If you receive flash burns, they should be treated by medical personnel.

A variety of special welder's clothing is used to protect parts of the body. The clothing selected varies with the size, location, and nature of the work to be performed. During any welding or cutting operation, you should always wear flameproof gauntlets. (SeeFig. 5.45) For gas welding and cutting, five-finger gloves like those shown in view A should be used. For electric arc welding, use the two-finger gloves (or mitts) shown in view B.

Fig. 5.45 Protective Gloves and Mitts

Both types of gloves protect the hands from heat and metal spatter. The two-finger gloves have an advantage over the five-finger gloves in that they reduce the danger of weld spatter and sparks lodging between the fingers. They also reduce finger chafing which sometimes occurs when five-finger gloves are worn for electric-arc welding.

Many light-gas welding and brazing jobs require no special protective clothing other than gloves and goggles. Even here, it is essential that you wear your work clothes properly. Sparks are very likely to lodge in rolled-up sleeves, pockets of clothing, or cuffs of trousers or overalls. Sleeves should be rolled down and the cuffs buttoned. The shirt collar, also, should be fully buttoned. Trousers should not be cuffed on the outside, and pockets not protected by button-down flaps should be eliminated from the front of overalls and aprons. All other clothing must be free of oil and grease. Wear high top-safety shoes; low-cut shoes are a hazard because sparks and molten metal could lodge in them, especially when you are sitting down.

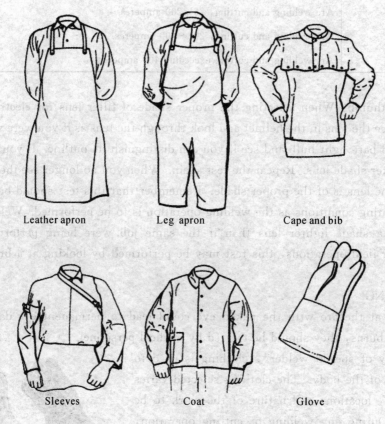

Fig. 5.47 Protective Clothing

Medium-gas and heavy-gas welding, all-electric welding, and welding in the vertical or overhead welding position require special flameproof clothing made of leather or other suitable material. This clothing is designed to protect you against radiated heat, splashes of hot metal, or sparks. This clothing consists of aprons, sleeves, combination sleeves and bib, jackets, and overalls. They afford a choice of protection depending upon the specific nature of the particular welding or cutting job. Sleeves provide satisfactory protection for welding operations at floor or bench level.

The cape and sleeves are particularly suited for overhead welding, because it protects the back of the neck, top of the shoulders, and the upperpart of the back and chest. Use of the bib, in combination with the cape and sleeves, gives added protection to the chest and abdomen. The jacket should be worn when there is a need for complete all-around protection to the upperpart of the body. This is especially true when several welders are working in close proximity to one another. Aprons and overalls provide protection to the legs and are suited for welding operations on the floor. Fig. 5.48 shows some of the protective clothing available to welders. To prevent head burns during overhead welding operations, you should wear leather or flameproof caps under the helmet. Earplugs also should be worn to keep sparks or splatter from entering and burning the ears.

Where the welder is exposed to falling or sharp objects, combination welding helmet/hard hats should be used. For very heavy work, fire-resistant leggings or high boots should be worn. Shoes or boots having exposed nailheads or rivets should NOT be worn. Oilskins or plastic clothing must NOT be worn in any welding operation.

NOTE: If leather protective clothing is not available, then woolen clothing is preferable to cotton. Woolen clothing is not as flammable as cotton and helps protect the operator from the changes in temperature caused by welding. Cotton clothing, if used, should be chemically treated to reduce its flammability.

Exercises

1. What is gas shielded arc welding?
2. List the protective clothes that you need when welding.
3. How to avoid distortion when welding?

Chapter 6 Lathes

Teaching Objectives

In this chapter, students are required to know the basic structure of lathe, lathe tools, tooling angles, operation procedures of lathe.

6.1 Brief Introduction

Lathe is a machine tool used mainly for shaping articles of metal (sometimes wood or other materials) by holding the workpiece and rotating the lathe while a tool bit advancing into the work for the cutting action (Fig. 6.1). Basically, a lathe can be used to produce screw threads, tapered work, to drill holes, knurl surfaces, and crankshafts. The typical lathe provides a variety of rotating speeds and a means to manually and automatically move the cutting tool into the workpiece.

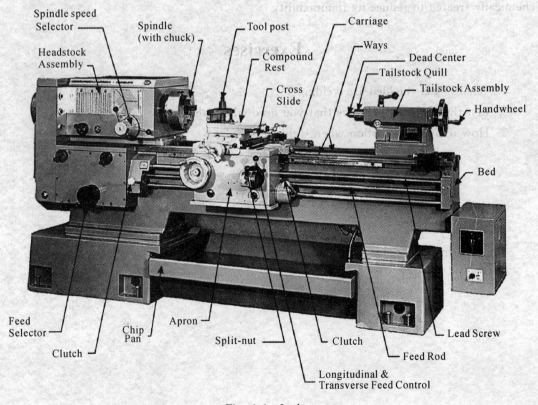

Fig. 6.1 Lathe

6.2 Types of Lathes

Lathes can be divided into three types: engine lathes, turret lathes, and special purpose lathes. Small lathes can be bench mounted, and can be transported in wheeled vehicles easily. The larger lathes are floor mounted and may require special transportation. A trained operator can accomplish more machining jobs with the engine lathe than with any other machine tool. Turret lathes and special purpose lathes are usually used in production or job shops for mass production of specialized parts. This chapter will be mainly about the various engine lathes.

6.3 Engine Lathes and Classification

Sizes

The size of an engine lathe is determined by the largest piece of stock that can be machined. Before machining workpiece, the following measurements must be considered: the diameter of the work that will swing over the bed and the length between lathe centers.

Categories

Slight differences in the various engine lathes make it easy to group them into three categories: lightweight bench engine lathes, precision tool room lathes, and gap lathes, also known as extension-type lathes. These lathe categories are shown in Fig. 6.2. Different manufacturers may use different lathe categories.

Lightweight

Lightweight bench engine lathes are generally small lathes with a swing of 10 inches or less, mounted to a bench or table top. These lathes can accomplish most machining jobs, but may be limited due to the size of the material that can be turned.

Precision

Precision tool room lathes are also known as standard manufacturing lathes, usually used for all lathe operations, such as turning, boring, drilling, reaming, producing screw threads, taper turning, knurling, and radius forming, and can be adapted for special milling operations with the appropriate fixture. This type of lathe can handle workplaces up to 25 inches in diameter and up to 200 inches long. However, the general size is about a 15-inch swing with 36-48 inches between centers. Many tool room lathes are used for special tool and die production due to the high accuracy of the machine.

Simplified Metal Works

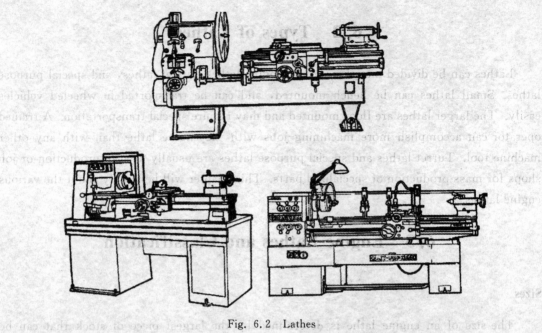

Fig. 6.2 Lathes

Gap Lathes

Gap or extension-type lathes are similar to toolroom lathes except that gap lathes can be adjusted to machine larger diameter and longer workplaces.

6.4 Lathe Components

Engine lathes all have the same general functional parts, even though the specific location or shape of a certain part may differ from one manufacturer. The bed is the foundation of the working parts of the lathe to another (Fig. 6.3).

The main feature of its construction are theways which are formed on its upper surface and run the full length of the bed. Ways provide the means for holding the tailstock and carriage, which slide along the ways, in alignment with the permanently attached headstock. The headstock is located on the operator's left end of the lathe bed. It contains the main spindle and oil reservoir and the gearing mechanism for obtaining various spindle speeds and for transmitting power to the feeding and threading mechanism. The headstock mechanism is driven by an electric motor connected either to a belt or pulley system or to a geared system. The main spindle is mounted on bearings in the headstock and is hardened and specially ground to fit different lathe holding devices. The spindle has a hole through its entire length to accommodate long workplaces. The hole in the nose of the spindle usually has a standard Morse taper which varies with the size of the lathe. Centers, collets, drill chucks, tapered shank drills and reamers may be inserted into the spindle. Chucks, drive plates, and faceplates may be screwed onto the spindle or clamped onto the spindle nose.

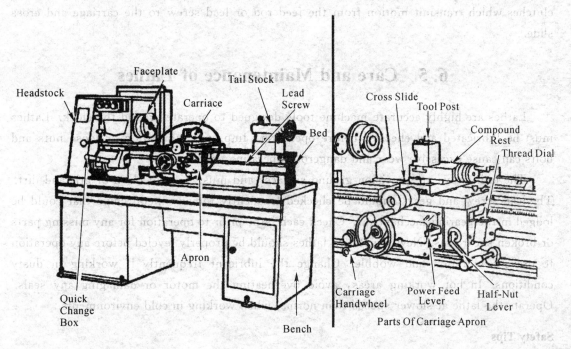

Fig. 6.3 Lathe Components

The tailstock is located on the opposite end of the lathe from the headstock. It supports one end of the work when machining between centers, supports long pieces held in the chuck, and holds various forms of cutting tools, such as drills, reamers, and taps. The tailstock is mounted on the ways and is designed to be clamped at any point along the ways. It has a sliding spindle that is operated by a hand wheel and clamped in position by means of a spindle clamp. The tailstock may be adjusted laterally (toward or away from the operator) by adjusting screws. It should be unclamped from the ways before any lateral adjustments are made, as this will allow the tailstock to be moved freely and prevent damage to the lateral adjustment screws.

The carriage includes the apron, saddle, compound rest, cross slide, tool post, and the cutting tool. It sits across the lathe ways and in front of the lathe bed. The function of the carriage is to carry and move the cutting tool. It can be moved by hand or by power and can be clamped into position with a locking nut. The saddle carries the cross slide and the compound rest. The cross slide is mounted on the dovetail ways on the top of the saddle and is moved back and forth at 90° to the axis of the lathe by the cross slide lead screw. The lead screw can be hand or power activated. A feed reversing lever, located on the carriage or headstock, can be used to cause the carriage and the cross slide to reverse the direction of travel. The compound rest is mounted on the cross slide and can be swiveled and clamped at any angle in a horizontal plane. The compound rest is used extensively in cutting steep tapers and angles for lathe centers. The cutting tool and tool holder are secured in the tool post which is mounted directly to the compound rest. The apron contains the gears and feed clutches which transmit motion from the feed rod or lead screw to the carriage and cross slide.

6.5 Care and Maintenance of Lathes

Lathes are highly accurate machine tools designed to operate around the clock. Lathes must be lubricated and checked before operation. Improper lubrication or loose nuts and bolts can cause excessive wear and dangerous operating conditions.

The lathe ways are precision ground surfaces and must be kept clean of grit and dirt. The lead screw and gears should be checked frequently for any metal chips that could be lodged in the gearing mechanisms. Check each lathe prior to operation for any missing parts or broken shear pins. Newly installed lathes should be properly leveled before any operation to prevent vibration and wobble. Change the lubricant frequently if working in dusty conditions. In hot working areas, avoid overheating the motor or damaging any seals. Operate the lathe at slower speeds than normal when working in cold environments.

Safety Tips

All lathe operators must be constantly aware of the safety hazards that are associated with using the lathe and must know all safety precautions to avoid accidents and injuries. Some important safety precautions to follow when using lathes are:

a) Correct dress is important, remove rings and watches, and roll sleeves above elbows.
b) Always stop the lathe before making adjustments.
c) Do not change spindle speeds until the lathe comes to a complete stop.
d) Handle sharp cutters, centers, and drills with care.
e) Remove chuck keys and wrenches before operating.
f) Always wear protective eye protection.
g) Handle heavy chucks with care and protect the lathe ways with a block of wood when installing a chuck.
h) Know where the emergency stop is before operating the lathe.
i) Use pliers or a brush to remove chips and swarf, never your hands.
j) Never lean on the lathe.
k) Never lay tools directly on the lathe ways. If a separate table is not available, use a wide board with a cleat on each side to lay on the ways.
l) Keep tools overhang as short as possible.
m) Never attempt to measure work while it is turning.
n) Never file lathe work unless the file has a handle.
o) File left-handed if possible.
p) Protect the lathe ways when grinding or filing.

q) Use two hands when sanding the workpiece. Do not wrap sand paper or emery cloth around the workpiece.

6.6 Tools and Equipment

6.6.1 General Purpose Cutting Tools

The lathe cutting tool or tool bit must be made of the correct material and ground to the correct angles to run efficiently. The most common tool bit is the general all-purpose bit made of high-speed steel. They are generally inexpensive, easy to grind on a bench or pedestal grinder, take lots of wear, and are strong enough for all-around repair and fabrication. High-speed steel tool bits can handle the high heat generated during cutting and are not changed after cooling. These tool bits are used for turning, facing, boring and other lathe operations. Tool bits made from special materials such as carbides, ceramics, diamonds, cast alloys are able to machine workplaces at very high speeds but are brittle and expensive for normal lathe work. High-speed steel tool bits are available in many shapes and sizes to accommodate any lathe operation.

6.6.2 Single Point Tool Bits

Single point tool bits can be one end of a high-speed steel tool bit or one edge of a carbide or ceramic cutting tool or insert. Basically, a single point cutter bit is a tool that has only one cutting action proceeding at a time. A machinist or machine operator should know the various terms applied to the single point tool bit to properly identify and grind different tool bits (Fig. 6.4).

Theshank is the main body of the tool bit. The nose is the part of the tool bit which is shaped to a point and forms the corner between the side cutting edge and the end cutting edge. The nose radius is the rounded end of the tool bit. The face is the top surface of the tool bit upon which the chips slide as they separate from the work piece. The side or flank of the tool bit is the surface just below and adjacent to the cutting edge. The cutting edge is the part of the tool bit that actually cuts into the workpiece, located behind the nose and adjacent to the side and face. The base is the bottom surface of the tool bit, which usually is ground flat during tool bit manufacturing. The end of the tool bit is the near-vertical surface which, with the side of the bit, forms the profile of the bit. The end is the trailing surface of the tool bit when cutting. The heel is the portion of the tool bit base immediately below and supporting the face.

6.6.3 Angles of Tool Bits

The successful operation of the lathe and the quality of work that may be achieved depend largely on the angles that form the cutting edge of the tool bit (Fig. 6.4). Most tools are hand ground to the desired shape on a bench or pedestal grinder. The cutting tool geometry for the rake and relief angles must be properly ground, but the overall shape of the tool bit is determined by the preference of the machinist or machine operator. Lathe tool bit shapes can be pointed, rounded, squared off, or irregular in shape and still cut quite well as long as the tool bit angles are properly ground for the type of material being machined. The angles are the side and back rake angles, the side and end cutting edge angles, and the side and end relief angles. Other angles to be considered are the radius on the end of the tool bit and the angle of the tool holder. After knowing how the angles affect the cutting action, some recommended cutting tool shapes can be considered.

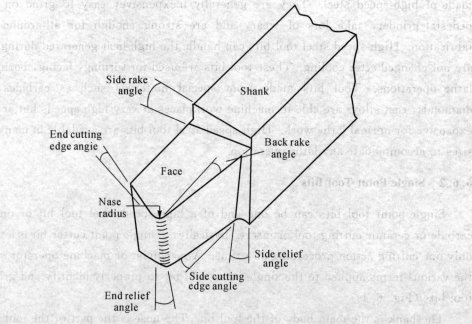

Fig. 6.4 Tool Bit Angle

Rake angle pertains to the top surface of the tool bit. There are two types of rake angles, the side and back rake angles (Fig. 6.4). The rake angle can be positive, negative, or have no rake angle at all. The tool holder can have an angle, known as the tool holder angle, which averages about 15°, depending on the model of tool holder selected. The tool holder angle combines with the back rake angle to provide clearance for the heel of the tool bit from the workpiece and to facilitate chip removal. The side rake angle is measured back from the cutting edge and can be a positive rake angle or have no rake at all.

Rake angles cannot be too great or the cutting edge will lose strength to support the cutting action. The side rake angle determines the type and size of chip produced during the cutting action and the direction that the chip travels when leaving the cutting tool. Chip breakers can be included in the side rake angle to ensure that the chips break up and do not become a safety hazard.

Side and relief angles, or clearance angles, are the angles formed behind and beneath the cutting edge that provide clearance or relief to the cutting action of the tool. There are two types of relief angles, side relief and end relief. Side relief is the angle ground into the tool bit, under the side of the cutting edge, to provide clearance in the direction of tool bit travel. End relief is the angle ground into the tool bit to provide front clearance to keep the tool bit heel from rubbing.

The end relief angle is supplemented by the tool holder angle and makes up the effective relief angle for the end of the tool bit.

Side and cutting edge angles are the angles formed by the cutting edge with the end of the tool bit (the end cutting edge angle), or with the side of the tool bit (the side cutting edge angle). The end cutting edge angle permits the nose of the tool bit to make contact with the work and aids in feeding the tool bit into the work. The side cutting edge angle reduces the pressure on the tool bit as it begins to cut. The side rake angle and the side relief angle combine to form the wedge angle (or lip angle) of the tool bit that provides for the cutting action.

A radius ground onto the nose of the tool bit can help strengthen the tool bit and provide for a smooth cutting action.

6.6.4 Shapes of Tool Bits

The overall shape of the lathe tool bits can be rounded, squared, or another shape as long as the proper angles are included Fig. 6.5. Tool bits are identified by the function they perform, such as turning or facing. They can also be identified as roughing tools or finishing tools. Generally, a roughing tool has a radius ground onto the nose of the tool bit that is smaller than the radius for a finishing or general purpose tool bit. Experienced machinists have found the following shapes to be useful for different lathe operations.

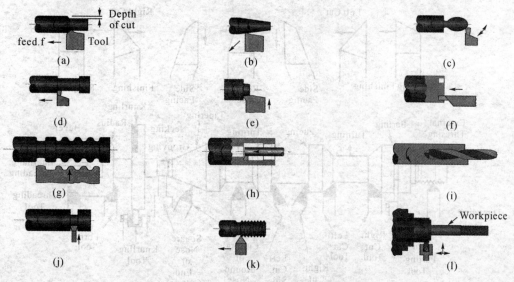

Fig. 6.5 Tool Bit Shape

(a) Straight Turning; (b) Taper Turning; (c) Profiling; (d) Turning and External Grooving; (e) Facing; (f) Face Grooving; (g) Cutting with a Form Tool; (h) Boring and Interanl Grooving; (i) Drilling; (j) Cutting off; (k) Threading; (l) Knurling

A right-hand turning tool bit is shaped to be fed from right to left. The cutting edge is on the left side of the tool bit and the face slopes down away from the cutting edge. The left side and end of the tool bit are ground with sufficient clearance to permit the cutting edge to bear upon the workpiece without the heel rubbing on the work. The right-hand turning tool bit is ideal for taking light roughing cuts as well as general all-around machining.

A left-hand turning tool bit is the opposite of the right-hand turning tool bit, designed to cut when fed from left to right. This tool bit is used mainly for machining close into a right shoulder. (Fig. 6.6)

The round-nose turning tool bit is very versatile and can be used to turn in either direction for roughing and finishing cuts. No side rake angle is ground into the top face when used to cut in either direction, but a small back rake angle may be needed for chip removal. The nose radius is usually ground in the shape of a half-circle with a diameter of about 1/32 inch.

The right-hand facing tool bit is intended for facing on right hand side shoulders and the right end of a workpiece. The cutting edge is on the left-hand side of the bit and the nose is ground very sharp for machining into a square corner. The direction of feed for this tool bit should be away from the center axis of the work, not going into the center axis.

A left-hand facing tool bit is the opposite of the right-hand facing tool bit and is intended to machine and face the left sides of shoulders.

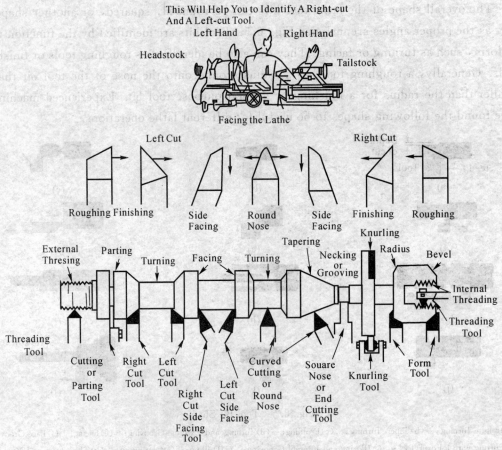

Fig. 6.6 Parting Tool Bit

6.6.5 Tool Bit and Lathe Operations (Fig. 6.7)

- Turning — produces straight, conical, curved, or grooved work pieces
- Facing — produces a flat surface at the end of the part
- Boring — to enlarge a hole
- Drilling — to produce a hole
- Cutting off — to cut off a work piece
- Threading — to produce threads
- Knurling — produces a regularly shaped roughness

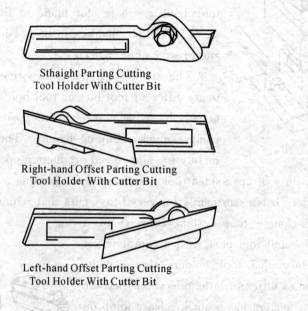

Sthaight Parting Cutting Tool Holder With Cutter Bit

Right-hand Offset Parting Cutting Tool Holder With Cutter Bit

Left-hand Offset Parting Cutting Tool Holder With Cutter Bit

Fig. 6.7

6.6.6 Tool Holders and Tool Posts (Fig. 6.8)

Lathe tool holders are designed to secure the tool bit at a fixed angle for properly machining a workpiece.

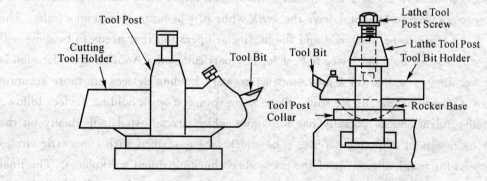

Fig. 6.8 Tool-holders and Round Tool Post

Simplified Metal Works

Tool holders are designed to work in conjunction with various lathe tool posts, onto which the tool holders are mounted. Tool holders for high speed steel tool bits come in various types. These tool holders are designed to be used with the standard round tool post. This tool post consists of the post screw, washer, collar, and rocker, and fits into the T-slot of the compound rest.

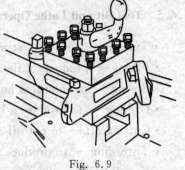
Fig. 6.9

Standard tool holders for high-speed steel cutting tools have a square slot made to fit a standard size tool bit shank. Tool bit shanks can be 1/4-inch, 5/16-inch, 3/8-inch, and greater (Fig. 6.9).

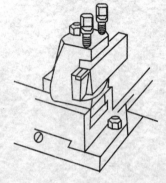
Fig. 6.10

The turret tool post is a swiveling block that can hold many different tool bits or tool holders. Each cutting tool can quickly be swiveled into cutting position and clamped into place using a quick clamping handle. The turret tool post is used mainly for high-speed production operations (Fig. 6.10).

The heavy-duty or open-sided tool post is used for holding a single carbide-tipped tool bit or tool holder. It is used mainly for very heavy cuts that require a rigid tool holder.

The quick-change tool system consists of a quick-change dovetail tool post with a complete set of matching dovetailed tool holders that can be quickly changed as different lathe operations become necessary. This system has a quick-release knob on the top of the tool post that allows tool changes in less than 5 seconds, which makes this system valuable for production machine shops (Fig. 6.11).

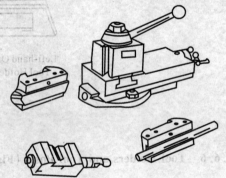
Fig. 6.11

WORK HOLDING DEVICES

Many different devices, such as chucks, collets, faceplates, drive plates, mandrels, and lathe centers, are used to hold and drive the work while it is being machined on a lathe. The size and type of work to be machined and the particular operation that needs to be done will determine which work holding device is best for any particular job. Another consideration is how much accuracy is needed for a job, since some work holding devices are more accurate than others. Operational details for some of the more common work holding devices follow.

The independent chuck, enerally has four jaws which are adjusted individually on the chuck face by means of adjusting screws. The chuck face is scribed with concentric circles which are used for rough alignment of the jaws when chucking round workpieces. The final adjustment is made by turning the workpiece slowly by hand and using a dial indicator to determine it's concentricity. The jaws are then readjusted as necessary to align the workpiece within the desired tolerances (Fig. 6.12).

The jaws of the independent chuck may be used as illustrated or may be reversed so that the steps face in the opposite direction; thus workpieces can be gripped either externally or internally. The independent chuck can be used to hold square, round, octagonal, or irregularly shaped workpieces in either a concentric or eccentric position due to the independent operation of each jaw.

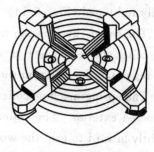

Fig. 6.12 Independant Chuck

Because of its versatility and capacity for fine adjustment, the independent chuck is commonly used for mounting odd-shaped workpieces which must be held with extreme accuracy.

The universal scroll chuck, usually has three jaws which move in unison as an adjusting pinion is rotated. The advantage of the universal scroll chuck is its ease of operation in centering work for concentric turning. This chuck is not as accurate as the independent chuck, but when in good condition it will center work within 0.002 to 0.003 inches of runout (Fig. 6.13).

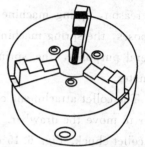

Fig. 6.13 Universal Scroll Chuck

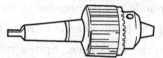

Fig. 6.14 Drill Chuck

The jaws are moved simultaneously within the chuck by a scroll or spiral-threaded plate. The jaws are threaded to the scroll and move an equal distance inward or outward as the scroll is rotated by the adjusting pinion. Since the jaws are individually aligned on the scroll, the jaws cannot usually be reversed. Some manufactures supply two sets of jaws, one for internal work and one for external work. Other manufactures make the jaws in two pieces so the outside, or gripping surface may be reversed. which can be interchanged.

The universal scroll chuck can be used to hold and automatically center round or hexagonal workpieces. Having only three jaws, the chuck cannot be used effectively to hold square, octagonal, or irregular shapes.

A combination chuck combines the features of the independent chuck and the universal scroll chuck and can have either three or four jaws. The jaws can be moved in unison on a scroll for automatic centering or can be moved individually if desired by separate adjusting screws.

The drill chuck, is a small universal chuck which can be used in either the headstock spindle or the tailstock for holding straight-shank drills, reamers, taps, or small diameter workpieces. The drill chuck has three or four hardened steel jaws which are moved together or apart by adjusting a tapered sleeve within which they are contained. The drill chuck is capable of centering tools and small-diameter workpieces to within 0.002 or 0.003 inch when firmly tightened (Fig. 6.14).

The collet chuck is the most accurate means of holding small workpieces in the lathe. The collet chuck consists of a spring machine collet and a collet attachment which secures and regulates the collet on the headstock spindle of the lathe.

The spring machine collet is a thin metal bushing with an accurately machined bore and a tapered exterior. The collet has three lengthwise slots to permit its sides being sprung slightly inward to grip the workpiece. To grip the workpiece accurately, the collet must be no more than 0.005 inch larger or smaller than the diameter of the piece to be chucked. For this reason, spring machine collets are available in increments of 1/64 inch. For general purposes, the spring machine collets are limited in capacity to 1 1/8 inch in diameter. For general purposes, the spring machine collets are limited in capacity to 1 1/8 inch in diameter.

The collet attachment consists of a collet sleeve, a drawbar, and a handwheel or hand lever to move the drawbar. The spring machine collet and collet attachment together form the collet chuck. Fig. 6.15 illustrates a typical collet chuck installation. The collet sleeve is fitted to the right end of the headstock spindle. The drawbar passes through the headstock spindle and is threaded to the spring machine collet. When the drawbar is rotated by means of the hand wheel, it draws the collet into the tapered adapter, causing the collet to tighten on the workpiece. Spring machine collets are available in different shapes to chuck square and hexagonal workpieces of small dimensions as well as round workpieces.

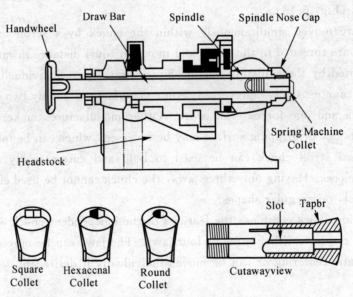

Fig. 6.15 The Spring Machine Collect Chuck
The collet chuck is the most accurate means of holding small workpieces

The Jacob's spindle-nose collet chuck is a special chuck is used for the Jacob's rubber flex collets. This chuck combines the functions of the standard collet chuck and drawbar into one single compact unit. The chuck housing has a handwheel on the outer diameter that turns to tighten or loosen the tapered spindle which holds the rubber flex collets. Rubber flex collets are comprised of devices made of hardened steel jaws in a solid rubber housing. These collets have a range of 1/8 inch per collet. The gripping power and accuracy remain constant throughout the entire collet capacity. Jacob's rubber flex collets are designed for heavy duty turning and possess two to four times the grip of the conventional split steel collet. The different sets of these collets are stored in steel boxes designed for holding the collets. Collets are normally stored in steel boxes designed for holding the collets (Fig. 6.16).

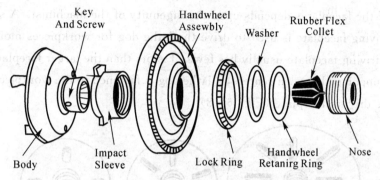

Fig. 6.16 The Jacob's Spindle-nose Collet Chuck

The step chuck, is a variation of the collet chuck, and it is intended for holding small round workpieces or discs for special machining jobs. Step chucks are blank when new, and then are machined in the lathe for an exact fit for the discs to be turned. The step chuck machine collet, which is split into three sections like the spring machine collet, is threaded to the drawbar of the collet attachment (Fig. 6.17).

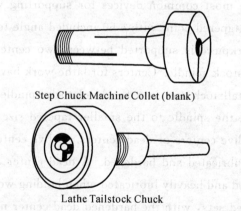

Fig. 6.17 Step Chuck and Tailstock Chuck

The lathe tailstock chuck, is a device designed to support the ends of workpieces in the tailstock when a lathe center cannot be used conveniently. The chuck has a taper arbor that fits into the lathe tailstock spindle. The three bronze self-centering jaws of the chuck will accurately close upon workpieces between 1/4 and 1 inch in diameter. The bronze jaws provide a good bearing surface for the workpiece. The jaws are adjusted to the diameter of the workpiece and then locked in place.

A lathe faceplate, is a flat, round plate that threads to the headstock spindle of the lathe. The faceplate is used for irregularly shaped workpieces that cannot be successfully held by chucks or mounted between centers. The workpiece is either attached to the faceplate using angle plates or brackets or bolted directly to the plate. Radial T-slots in the faceplate surface facilitate mounting workpieces. The faceplate is valuable for mounting workpieces in which an eccentric hole or projection is to be machined. The number of applications of the faceplates depends upon the ingenuity of the machinist. A small faceplate known as a driving faceplate is used to drive the lathe dog for workpieces mounted between centers. The driving faceplate usually has fewer T-slots than the larger faceplates. When the workpiece is supported between centers, a lathe dog is fastened to the workpiece and engaged in a slot of the driving faceplate (Fig. 6.18).

Large Sldtted   Small Slotted   Driving
                 Face Plate   Face Plate

Fig. 6.18 Face Plate

Lathe centers are the most common devices for supporting workpieces in the lathe. Most lathe centers have a tapered point with a 60 included angle to fit workpiece holes with the same angle. The workpiece is supported between two centers, one in the headstock spindle and one in the tailstock spindle. Centers for lathe work have standard tapered shanks that fit directly into the tailstock and into the headstock spindle using a center sleeve to convert the larger bore of the spindle to the smaller tapered size of the lathe center. The centers are referred to as live centers or dead centers. A live center revolves with the work and does not need to be lubricated and hardened. A dead center does not revolve with the work and must be hardened and heavily lubricated when holding work. Live and dead centers commonly come in matched sets, with the hardened dead center marked with a groove near the conical end point (Fig. 6.19).

The ball bearing live center is a special center mounted in a ball bearing housing that lets the center turn with the work and eliminates the need for a heavily lubricated dead center. Ball bearing types of centers can have interchangeable points which make this center a versatile tool in all lathe operations. Modern centers of this type can be very accurate.

Descriptions for some common lathe centers follow.

The male center or plain center is used in pairs for most general lathe turning operations. The point is ground to a 60° cone angle. When used in the headstock spindle where it revolves with the workpiece, it is commonly called a live center. When used in the tailstock spindle where it remains stationary when the workpiece is turned, it is called a dead center. Dead centers are always made of hardened steel and must be lubricated very often to prevent overheating.

The half male center is a male center that has a portion of the cone cut away. The half male center is used as a dead center in the tailstock where facing is to be performed. The cutaway portion of the center faces the cutting tool and provides the necessary clearance for the tool when facing the surface immediately around the drilled center in the workpiece.

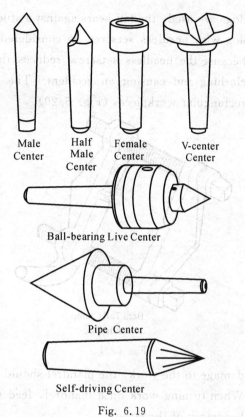

Fig. 6.19

The V-center is used to support round workpieces at right angles to the lathe axis for special operations such as drilling or reaming. The pipe center is similar to the male center but its cone is ground to a greater angle and is larger in size. It is used for holding pipe and tubing in the lathe. The female center is conically bored at the tip and is used to support workpieces that are pointed on the end. A self-driving lathe center is a center with serrated ground sides that can grip the work while turning between centers without having to use lathe dogs.

Lathe dogs are cast metal devices used to provide a firm connection between the headstock spindle and the workpiece mounted between centers. This firm connection permits the workpiece to be driven at the same speed as the spindle under the strain of cutting. Lathe dogs may have bent tails or straight tails. When bent-tail dogs are used, the tail fits into a slot of the driving faceplate. When straight-tail

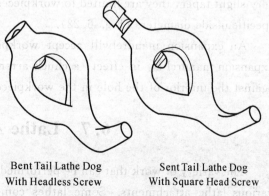

Bent Tail Lathe Dog
With Headless Screw

Sent Tail Lathe Dog
With Square Head Screw

Fig. 6.20

dogs are used, the tail bears against a stud projecting from the faceplate. The bent-tail lathe dog with headless setscrew is considered safer than the dog with the square head screw because the headless setscrew reduces the danger of the dog catching in the operator's clothing and causing an accident. The bent-tail clamp lathe dog is used primarily for rectangular workpieces (Fig. 6.20).

MANDRELS

A workpiece which cannot be held between centers because its axis has been drilled or bored, and which is not suitable for holding in a chuck or against a faceplate, is usually machined on a mandrel. A mandrel is a tapered axle pressed into the bore of the workpiece to support it between centers (Fig. 6.21).

A mandrel should not be confused with an arbor, which is a similar device but used for holding tools rather than workpieces. To prevent damage to the work, the mandrel should always be oiled before being forced into the hole. When turning work on a mandrel, feed toward the large end which should be nearest the headstock of the lathe.

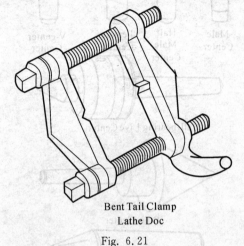

Bent Tail Clamp
Lathe Doc
Fig. 6.21

A solid machine mandrel is generally made from hardened steel and ground to a slight taper. It has very accurately countersunk centers at each end for mounting between centers. The ends of the mandrel are smaller than the body and have machined flats for the lathe dog to grip. The size of the solid machine mandrel is always stamped on the large end of the taper. Since solid machine mandrels have a very slight taper, they are limited to workpieces with specific inside diameters (Fig. 6.22).

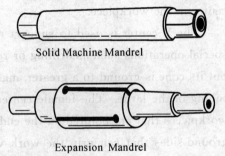

Solid Machine Mandrel

Expansion Mandrel
Fig. 6.22

An expansion mandrelwill accept workpieces having a greater range of sizes. The expansion mandrel is, in effect, a chuck arranged so that the grips can be forced outward against the interior of the hole in the workpiece.

6.7 Lathe Attachments

The variety of work that can be performed on the lathe is greatly increased by the use of various lathe attachments. Some lathes come equipped with special attachments; some attachments must be ordered separately. Some common lathe attachments are the steady rest with cathead, the follower rest, the tool post grinding machine, the lathe micrometer stop, the lathe milling fixture, the lathe coolant attachment, the lathe indexing fixture, and the milling-grinding-drilling-slotting attachment (or Versa-Mil). The lathe indexing fixture and Versa-Mil unit are detailed in Chapter 9. Descriptions for the other lathe attachments follows (Fig. 6.23).

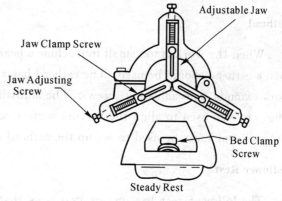

Fig. 6.23

Rests

Workpieces often need extra support, especially long, thin workpieces that tend to spring away from the tool bit. Three common supports or rests are the steady rest, the cathead, and the follower rest (Fig. 6.24).

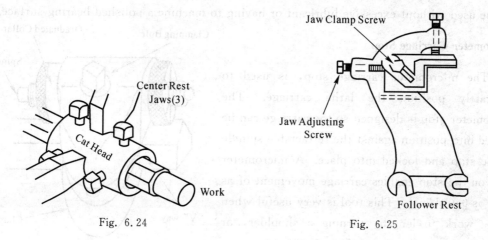

Fig. 6.24

Fig. 6.25

Steady Rest

The steady rest, also called a center rest, is used to support long workpieces for turning and boring operations. It is also used for internal threading operations where the workpiece projects a considerable distance from the chuck or faceplate. The steady rest is clamped to the lathe bed at the desired location and supports the workpiece within three adjustable jaws. The workpiece must be machined with a concentric bearing surface at the point where the steady rest is to be applied. The jaws must be carefully adjusted for proper alignment and locked in position. The area of contact must be lubricated frequently. The top section of the steady rest swings away from the bottom section to permit removal of the workpiece without disturbing the jaw setting (Fig. 6.25).

Cathead

When the work is too small to machine a bearing surface for the adjustable jaws to hold, then a cathead should be used. The cathead has a bearing surface, a hole through which the work extends, and adjusting screws. The adjusting screws fasten the cathead to the work. They are also used to align the bearing surface so that it is concentric to the work axis. A dial indicator must be used to set up the cathead to be concentric and accurate.

Follower Rest

The follower rest has one or two jaws that bear against the workpiece. The rest is fastened to the lathe carriage so that it will follow the tool bit and bear upon the portion of the workpiece that has just been turned. The cut must first be started and continued for a short longitudinal distance before the follower rest may be applied. The rest is generally used only for straight turning and for threading long, thin workpieces. Steady rests and follower rests can be equipped with ball-bearing surfaces on the adjustable jaws. These types of rests can be used without excessive lubricant or having to machine a polished bearing surface.

Micrometer Carriage Stop

The micrometer carriage stop, is used to accurately position the lathe carriage. The micrometer stop is designed so the carriage can be moved into position against the retractable spindle of the stop and locked into place. A micrometer gage on the stop enables carriage movement of as little as 0.001 inch. This tool is very useful when facing work to length, turning a shoulder, or cutting an accurate groove (Fig. 6.26).

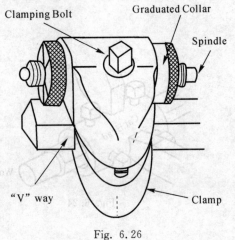

Fig. 6.26

6.8 Tools Necessary for Lathe Work

In order to properly setup and operate most engine lathes, it is recommended to have the following tools on hand. A machinist tool box with all wrenches, screwdrivers, and common hand tools. A dial indicator may be necessary for some procedures on the lathe. References, charts, tables, and other predetermined data on machine operations may be useful to lathe operators. Keep all safety equipment, along with necessary cleaning marking, and lubricating equipment, in the immediate lathe area to use as needed.

Cutting Fluids

The purposes of using cutting fluids on the lathe are to cool the tool bit and workpiece that are being machined, increase the life of the cutting tool, make a smoother surface finish, deter rust, and wash away chips. Cutting fluids can be sprayed, dripped, wiped, or flooded onto the point where the cutting action is taking place. Generally, cutting fluids should only be used if the speed or cutting action requires the use of cutting fluids. Descriptions of some common cutting fluids used on the lathe follow.

Lard Oil

Pure lard oil is one of the oldest and best cutting oils. It is especially good for thread cutting, tapping, deep hole drilling, and reaming. Lard oil has a high degree of adhesion or oiliness, a relatively high specific heat, and its fluidity changes only slightly with temperature. It is an excellent rust preventive and produces a smooth finish on the workpiece. Because lard oil is expensive, it is seldom used in a pure state but is combined with other ingredients to form good cutting oil mixtures.

Mineral Oil

Mineral oils are petroleum-base oils that range in viscosity from kerosene to light paraffin oils. Mineral oil is very stable and does not develop disagreeable odors like lard oil; however, it lacks some of the good qualities of lard oil such as adhesion, oiliness, and high specific heat. Because it is relatively inexpensive, it is commonly mixed with lard oil or other chemicals to provide cutting oils with desirable characteristics. Two mineral oils, kerosene and turpentine, are often used alone for machining aluminum and magnesium. Paraffin oil is used alone or with lard oil for machining copper and brass.

Mineral-lard Cutting Oil Mixture

Various mixtures of mineral oils and lard oil are used to make cutting oils which combine the good points of both ingredients but prove more economical and often as effective as pure lard oil.

Sulfurized Fatty-mineral Oil

Most good cutting oils contain mineral oil and lard oil with various amounts of sulfur and chlorine which give the oils good antiweld properties and promote free machining. These oils play an important part in present-day machining because they provide good finishes on most materials and aid the cutting of tough material.

Soluble Cutting Oils

Water is an excellent cooling medium but has little lubricating value and hastens rust and corrosion. Therefore, mineral oils or lard oils which can be mixed with water are often used to form a cutting oil. A soluble oil and water mix has lubricating qualities dependent upon the strength of the solution. Generally, soluble oil and water is used for rough cutting where quick dissipation of heat is most important. Borax and trisodium phosphate (TSP) are sometimes added to the solution to improve its corrosion resistance.

Soda-water Mixtures

Salts such as soda ash and TSP are sometimes added to water to help control rust. This mixture is the cheapest of all coolants and has practically no lubricating value. Lard oil and soap in small quantities are sometimes added to the mixture to improve its lubricating qualities. Generally, soda water is used only where cooling is the prime consideration and lubrication a secondary consideration. It is especially suitable in reaming and threading operations on cast iron where a better finish is desired.

White Lead and Lard Oil Mixture

White lead can be mixed with either lard oil or mineral oil to form a cutting oil which is especially suitable for difficult machining of very hard metals.

6.9 Basic Lathe Operations

The lathe is capable of a huge range of operations, especially when combined with additional accessories and tooling. However there are 3 basic operations which cover many of the tasks done on the lathe.

Turning to a required diameter.
Drilling a hole using the tailstock.
Turning a bar to length.

Turning a Bar to Length

In this example a 25mm off-cut of 12mm diameter aluminium bar, was turned to an exact length of 20mm.

First, the bar was measured to establish the current length (Fig. 6.27).

Before measuring, both ends of the bar were faced, so that it could be measured accurately.

The 3 jaw chuck was used to hold the work-piece and each end was lightly faced to leave them both parallel (Fig. 6.28).

The Bar length was measured using a caliper and was 23.38mm.

Current length−Required length=Material to be removed.

Therefore : 23.38mm−20.00mm=3.38mm to be removed

To remove this amount of material the work-piece was placed back in the chuck and the top-slide wound to zero.

Next, the saddle was wound towards the work-piece until the tool just touched. This had to be done carefully.

Fig. 6.27   Fig. 6.28

The top slide was incremented in 10 digit steps and after 150 divisions, the work piece was removed from the chuck to check the length. This was done because the amount to be removed was quite high and resetting the cut from near to the final length would remove any built up error.

This re-measurement showed that an error of 0.03mm had accumulated (Fig. 6.29).

Fig. 6.29

Turning to a Diameter.

As with turning to length, when turning a diameter, first the starting diameter had to be found.

The raw bar could not be assumed to be perfectly round, nor the chuck completely true and so a light cut was taken to give a datum surface.

The plan was to turn to a diameter of 20mm in this example (Fig. 6.30).

Fig. 6.30

This datum cut was taken with the cross slide dial set to zero, to make subsequent cuts easier.

The starting diameter was measured and the required diameter subtracted to determine the cut to be taken.

Current Diameter − Required Diameter = Amount of material to be removed.

21.77mm − 20.00mm = 1.77mm to be removed

Each division on the cross-slide was equal to 0.04mm off the diameter (Fig. 6.31).

So the required cut was a total of 1.77mm/0.04 = 44.25 divisions.

Fig. 6.31

The cross-slide was then incremented in steps of 10 divisions to get close to the final diameter.

Periodic checks were made and the amount to be removed recalculated, as the component neared final size. Rechecking was easier than when turning to length because the material could remain in the chuck.

The lathe power feed was used to give the best surface finish in the final cut (Fig. 6.32).

Fig. 6.32

Drilling with the Tailstock.

The tailstock can be used to drill holes in the center-line of the lathe. In this example a 5mm through hole was drilled in a brass off-cut.

First the face of the work-piece was cut square to prevent the drill from wandering or slipping offcenter (Fig. 6.33).

Chapter 6　Lathes

Fig. 6.33

A centre drill was used to accurately mark the centre.

The centre drill was used to a depth deep enough to fully guide the tip of the next drill to be used. For this 5mm hole, the next drill to be used was 4.5mm diameter (Fig. 6.34).

Fig. 6.34

Drills were used in progressively larger sizes up to 0.1mm below the final dimension, in this case 4.9mm was the last drill used (Fig. 6.35).

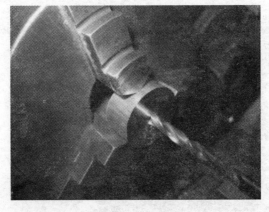

Fig. 6.35

The hole was finished to final size using a reamer. The lathe was used on the lowest open speed or highest back gear speed for this operation (Fig. 6.36).

Fig. 6.36

Exercises

1. Reread the safety tips for lathe work.
2. What is profiling operation?
3. What is facing operation?

Chapter 7 Planing Machines

Teaching Objectives

After studying this chapter, you should be able to:
- Identify various planing machines.
- Explaining how planing machines operate.
- Describe the industrial applications of planing machines.

7.1 Type of Planing Machines

Planing machines are designed to machine horizontal, vertical, and/or angular planed flat surfaces. These machines are classified into several categories.

7.1.1 Shaper

The shaper has a single point cutting tool that moves back and forth over the work. Being too slow for modern production techniques, most work done on a shaper is now performed by other machine tools. However, the shaper is still found in some specialty machine shops because it is easily and inexpensively tooled for some one-of-a-kind jobs (Fig. 7.1).

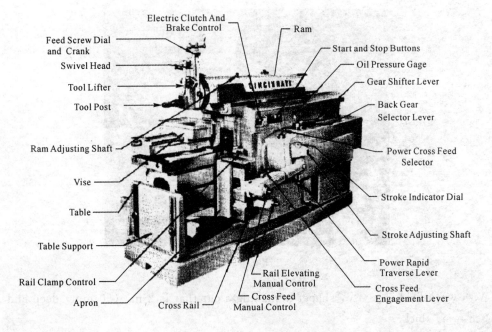

Fig. 7.1 Structure of a Shaper

On a shaper, the work is stationary and the single point cutting tool moves against it. While primarily used to machine flat surfaces, a skillful machinist can manipulate it to cut curved and irregular shapes, slots, grooves, and keyways. Work is usually mounted in a vise. As with any machine tool, carefully examine a shaper to be sure it is in safe operating condition (Fig. 7.2).

The machine should also be lubricated according to the manufacturer's specifications.

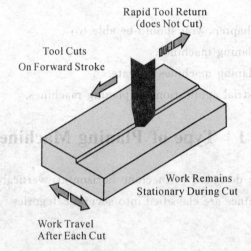

Fig. 7.2

SAFETY NOTE!

Never attempt to operate these machines while your senses are impaired by medication or other substances (Fig. 7.3).

Fig. 7.3

With work held in a vise, shaper is making a cut that is 2 in. (50 mm) deep and 1/32 in. (0.8 mm) thick.

Shaper size is determined by maximum work length that can be machined in one pass (Fig. 7.4).

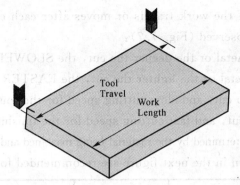

Fig. 7.4

Mounting work on shaper

The position of the vise is important when using a shaper. It should be positioned so the machining can be done in the shortest possible time (Fig. 7.5).

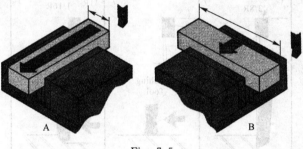

Fig. 7.5

Assuming that the cutter is making the same number of strokes per minute, the setup shown in A will permit the work to be done in about a third of the time needed to machine the work in the setup B.

The cutting stroke of the machine should be adjusted to work as shown in the left Figure. Make the adjustments as recommended by the machine's manufacturer.

The 1/4 in. (6.5 mm) allowance at end of stroke provides ample chip clearance, while 1/2 in. (12.5 mm) allowance permits cutter to drop back into cutting position for next stroke (Fig. 7.6).

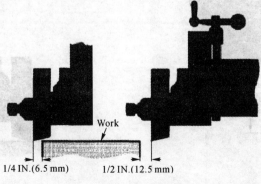

Fig. 7.6

— 105 —

Shaper cutting speed and feed

The shaper speed is the number of cutting strokes the ram makes per minute. The shaper feed is the distance the work travels or moves after each cutting stroke. Generally, the following should be observed (Fig. 7.7):

(1) The harder the metal or the deeper the cut, the SLOWER the cutting speed.

(2) The softer the metal or the lighter the cut, the FASTER the cutting speed.

(3) Coarse feed, deep cut, and slow cutting speed for the roughing cut.

(4) Fine feed, light cut, and fast cutting speed for the finishing cut.

Cutting tool shape is determined by the material being machined and the degree of finish desired. The tool shapes shown in the next figure are recommended for mild steel.

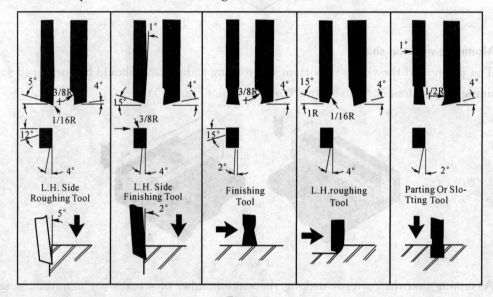

Fig. 7.7

Avoid excessive overhang of slide and/or cutting tool.

Excessive overhang causes chatter, producing a rough finish.

Keep slide up and a short grip on tool for increased rigidity (Fig. 7.8).

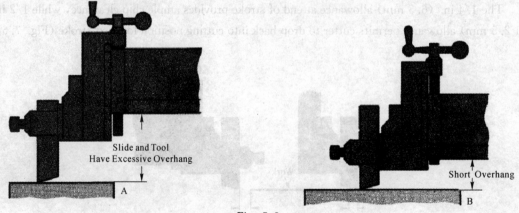

Fig. 7.8

7.1.2 Slotter

The chief difference between a shaper and a slotter is the direction of the cutting action. The slotting machine is classified as a VERTICAL SHAPER. It is used to cut slots, keyways (both internal and external), and to machine internal and external gears (Fig. 7.9).

Fig. 7.9

Vertical gear shaper is cutting a 120 in. (3050 mm) external gear (Fig. 7.10).

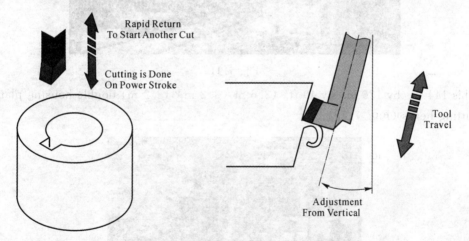

Fig. 7.10

Slotter, unlike shaper, moves vertically rather than horizontally and work is held stationary. Cutting head of a slotter can also be adjusted to make angular cuts.

7.1.3 Planer

The planer differs from the shaper in that the WORK travels back and forth while the cutter remains stationary. A planer can handle work that is too large to be machined on most milling machines. Planers are large pieces of equipment. Some are large enough to handle and machine work up to 20ft. (6.1 m) wide and twice as long (Fig. 7.11).

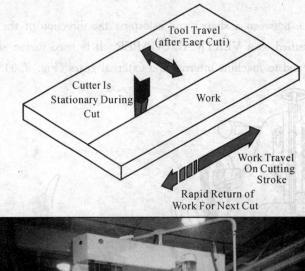

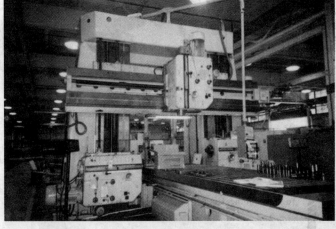

Fig. 7.11

This 144 in. by 126 in. by 40 ft. (3.6 m×3.2 m×12.2 m) double housing planer has two cutting heads (Fig. 7.12).

Fig. 7.12

This is a 42 in. by 42 in. by 10ft. (1.05 m by 1.05 m by 3.0 m) open side hydraulic planer.

7.2 Broaching

Broaching is similar to shaping, but instead of a single cutting tool advancing slightly after each stroke, the broach is a long tool with MANY teeth. Broaching machines are designed to push or pull this multi-tooth cutting tool across the work (Fig. 7.13).

The right diagram shows how a broach operates. A multitooth cutting tool moves against work. Operation may be on a vertical or horizontal plane.

Each tooth on the broach (cutting tool) removes only a small portion of the material being machined (Fig. 7.14).

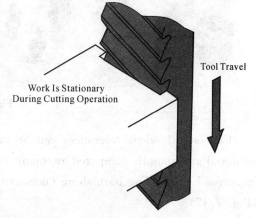

Fig. 7.13

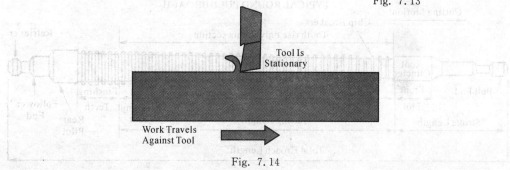

Fig. 7.14

Roughing, semi-finishing, and finishing teeth are usually on the same broach. The next drawing shows a greatly shortened section of internal broaching tool and a cross section of splines it cuts. Pilot guides cutter in work. Each cutting tooth increases slightly in size until specified size is attained. The machining operation can be completed in a single pass (Fig. 7.15).

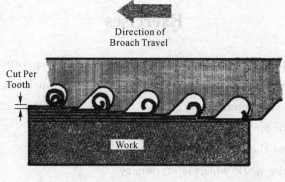

Fig. 7.15

Simplified Metal Works

When properly employed, broaching can remove material faster than any other machining technique. Small parts can often be stacked and shaped in a single pass of the tool. Larger units, such as auto engine blocks, may require several passes to machine all surfaces of the part (Fig. 7.16).

Fig. 7.16

Consistently close tolerances can be maintained by broaching. While surface finishes produced are smooth compared to many other machining processes, they can be further improved by providing burnishing (noncutting) elements to the finishing end of the broach (Fig. 7.17).

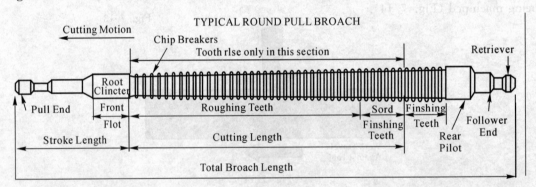

Fig. 7.17

Broaching machines are available in vertical and horizontal configurations and in many sizes. Some are fitted with dual rams. Work requiring multiple broaching passes can be transferred from one ram to the other to reduce handling time.

Exercises

Please do not write in the text. Place your answers on a separate sheet of paper.
1. The shaper is a machine used to machine _____ surfaces.
2. How is shaper size determined?
3. The cutting tool on the shaper:
a. Is stationary and the work moves against it.
b. Moves across the work which is stationary.

Chapter 7 Planning Machines

    c. Is moved across work which, in turn, moves at a slower speed in opposite direction.

    d. All of the above.

    e. None of the above.

4. With a harder metal or a deeper cut, the cutting speed should be _____

5. With softer metal or a lighter cut, the cutting speed should be _____

6. Use feed, cut, and _____ cutting speed for the roughing cut.

7. Use feed, cut, and _____ cutting speed for the finishing cut.

8. The vertical shaper is also known as a _____.

9. When is a planer needed to machine work?

10. The cutting tool on a planer:

    a. Is stationary and the work moves against it.

    b. Moves across the work which is stationary.

    c. Is pulled or pushed across the work.

    d. All of the above.

    e. None of the above.

11. How does broaching differ from other planing machines?

12. What is unique about the cutting tool used on a broaching machine?

Chapter 8 Milling and Milling Machines

Teaching Objectives

In this chapter, students should get familiar with various milling machines, major parts of milling machines, how to mount cutters, and various milling operations.

8.1 Introduction

Milling machines were first invented and developed by Eli Whitney to mass produce interchangeable musket parts. Although crude, these machines assisted man in maintaining accuracy and uniformity while duplicating parts that could not be manufactured with the use of a file. Development and improvements of the milling machine and components continued, which resulted in the manufacturing of heavier arbors and high speed steel and carbide cutters. These components allowed the operator to remove metal faster, and with more accuracy, than previous machines. Variations of milling machines were also developed to perform special milling operations. During this era, computerized machines have been developed to alleviate errors and provide better quality in the finished product.

8.2 Milling Machines

The milling machine removes metal with a revolving cutting tool called a milling cutter. With various attachments, milling machines can be used for boring, slotting, circular milling dividing, and drilling. This machine can also be used for cutting keyways, racks and gears and for fluting taps and reamers.

Milling machines are basically classified as being horizontal or vertical to indicate the axis of the milling machine spindle. These machines are also classified as knee-type, ram-type, manufacturing or bed type, and planer-type milling machines. Most machines have self-contained electric drive motors, coolant systems, variable spindle speeds, and power operated table feeds.

8.2.1 Knee-type Milling Machines

Knee-type milling machines are characterized by a vertical adjustable worktable resting on a saddle supported by a knee. The knee is a massive casting that rides vertically on the milling machine column and can be clamped rigidly to the column in a position where the milling head and the milling machine spindle are properly adjusted vertically for operation.

Chapter 8 Milling and Milling Machines

Floor-mounted Plain Horizontal Milling Machine

The floor-mounted plain horizontal milling machine's column contains the drive motor, gearing and a fixed-position horizontal milling machine spindle. An adjustable overhead arm, containing one or more arbor supports, projects forward from the top of the column. The arm and arbor supports are used to stabilize long arbors, upon which the milling cutters are fixed. The arbor supports can be moved along the overhead arm to support the arbor wherever support is desired. This support will depend on the location of the milling cutter or cutters on the arbor. The knee of the machine rides up or down the column on a rigid track. A heavy, vertical positioned screw beneath the knee is used for raising and lowering. The saddle rests upon the knee and supports the worktable. The saddle moves in and out on a dovetail to control the cross-feed of the worktable. The worktable traverses to the right or left upon the saddle, feeding the workpiece past the milling cutter. The table may be manually controlled or power fed.

Bench-type Plain Horizontal Milling Machine

The bench-type plain horizontal milling machine is a small version of the floor-mounted plain horizontal milling machine; it is mounted to a bench or a pedestal instead of directly to the floor. The milling machine spindle is horizontal and fixed in position. An adjustable overhead arm and support are provided. The worktable is generally not power fed on this size machine. The saddle slides on a dovetail on the knee providing cross-feed adjustment. The knee moves vertically up or down the column to position the worktable in relation to the spindle.

Floor-mounted Universal Horizontal Milling Machine

The basic difference between a universal horizontal milling machine and a plain horizontal milling machine is in the adjustment of the worktable, and in the number of attachments and accessories available for performing various special milling operations. The universal horizontal milling machine has a worktable that can swivel on the saddle with respect to the axis of the milling machine spindle, permitting workpieces to be adjusted in relation to the milling cutter.

The universal horizontal milling machine also differs from the plain horizontal milling machine in that it is of the ram type; i.e., the milling machine spindle is in a swivel cutter head mounted on a ram at the top of the column. The ram can be moved in or out to provide different positions for milling operations.

8.2.2 Ram-type Milling Machines

Description

The ram-type milling machine is characterized by a spindle mounted to a movable housing on the column, permitting positioning the milling cutter forward or rearward in a horizontal plane. Two widely used ram-type milling machines are the floor-mounted universal milling machine and the swivel cutter head ram-type milling machine.

Swivel Cutter Head Ram-type Milling Machine

A cutter head containing the milling machine spindle is attached to the ram. The cutter head can be swiveled from a vertical to a horizontal spindle position, or can be fixed at any desired angular position between the vertical and horizontal. The saddle and knee are driven for vertical and crossfeed adjustment; the worktable can be either hand driven or power driven at the operator's choice.

8.3 Major Components of Milling Machines

The machinist must know the name and purpose of each of the main parts of a milling machine to understand the operations discussed in this text. Keep in mind that although we are discussing a knee and a column milling machine, this information can be applied to other types. Use Fig. 8.1 (which illustrates a plain knee and column milling machine) to help become familiar with the location of the various parts of these machines.

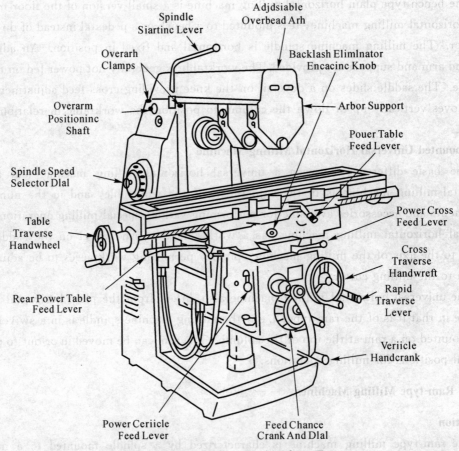

Fig. 8.1 Plain Milling Machine-knee Type

(1) Column. The column, including the base, is the main casting which supports all

other parts of the machine. An oil reservoir and a pump in the column keeps the spindle lubricated. The column rests on a base that contains a coolant reservoir and a pump that can be used when performing any machining operation that requires a coolant (Fig. 8.2).

(2) Knee. The knee is the casting that supports the table and the saddle. The feed change gearing is enclosed within the knee. It is supported and can be adjusted by the elevating screw. The knee is fastened to the column by dovetail ways. The lever can be raised or lowered either by hand or power feed. The hand feed is usually used to take the depth of cut or to position the work, and the power feed to move the work during the machining operation.

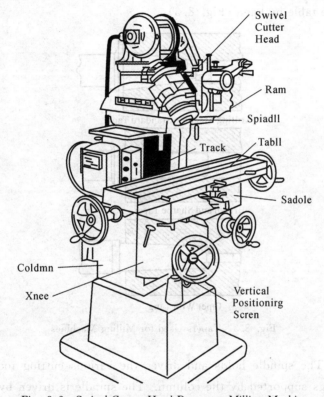

Fig. 8.2 Swivel Cutter Head Ram-type Milling Machine

(3) Saddle and swivel table. The saddle slides on a horizontal dovetail, parallel to the axis of the spindle, on the knee. The swivel table (on universal machines only) is attached to the saddle and can be swiveled approximately 45° in either direction.

(4) Power feed mechanism. The power feed mechanism is contained in the knee and controls the longitudinal, transverse (in and out) and vertical feeds. The desired rate of feed can be obtained on the machine by positioning the feed selection levers as indicated on the feed selection plates. On some universal knee and column milling machines the feed is obtained by turning the speed selection handle until the desired rate of feed is indicated on the feed dial. Most milling machines have a rapid traverse lever that can be engaged when a temporary increase in speed of the longitudinal, transverse, or vertical feeds is required. For

example, this lever would be engaged when positioning or aligning the work.

For safety reasons, extreme caution should be exercised while using the rapid traverse controls.

(5) Table. The table is the rectangular casting located on top of the saddle. It contains several T-slots for fastening the work or workholding devices. The table can be moved by hand or by power. To move the table by hand, engage and turn the longitudinal hand crank. To move it by power, engage the longitudinal directional feed control lever. The longitudinal directional control lever can be positioned to the left, to the right, or in the center. Place the end of the directional feed control lever to the left to feed the table to the left. Place it to the right to feed the table to the right. Place it in the center position to disengage the power feed, or to feed the table by hand (Fig. 8.3).

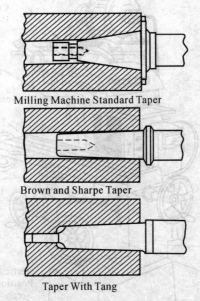

Fig. 8.3 Tapers Used for Milling Machines

(6) Spindle. The spindle holds and drives the various cutting tools. It is a shaft, mounted on bearings supported by the column. The spindle is driven by an electric motor through a train of gears, all mounted within the column. The front end of the spindle, which is near the table, has an internal taper machined on it. The internal taper (3 1/2 inches per foot) permits mounting tapered-shank cutter holders and cutter arbors. Two keys, located on the face of the spindle, provide a positive drive for the cutter holder, or arbor. The holder or arbor is secured in the spindle by a drawbolt and jamnut, as shown in Fig. 8.3. Large face mills are sometimes mounted directly to the spindle nose.

(7) Overarm. The overarm is the horizontal beam to which the arbor support is fastened. The overarm, may be a single casting that slides in the dovetail ways on the top of the column. It may consist of one or two cylindrical bars that slide through the holes in the column. On some machines to position the overarm, first unclamp the locknuts and then extend the overarm by turning a crank. On others, the overarm is moved by merely pushing

Chapter 8 Milling and Milling Machines

on it. The overarm should only be extended far enough to so position the arbor support as to make the setup as rigid as possible. To place the arbor supports on an overarm, extend one of the bars approximately 1-inch farther than the other bar. Always tighten the locknuts after the overarm is positioned. On some milling machines, the coolant supply nozzle is fastened to the overarm. The nozzle can be mounted with a split clamp to the overarm after the arbor support has been placed in position.

(8) Arbor Support. The arbor support is a casting containing a bearing which aligns the outer end of the arbor with the spindle. This helps to keep the arbor from springing during cutting operations. Two types of arbor supports are commonly used. One type has a small diameter bearing hole, usually 1-inch maximum in diameter. The other type has a large diameter bearing hole, usually up to 2 3/4 inches. An oil reservoir in the arbor support keeps the bearing surfaces lubricated. An arbor support can be clamped anywhere on the overarm. Small arbor supports give additional clearance below the arbor supports when small diameter cutters are being used. Small arbor supports can provide support only at the extreme end of the arbor, for this reason they are not recommended for general use. Large arbor supports can be positioned at any point on the arbor. Therefore they can provide support near the cutter, if necessary. The large arbor support should be positioned as close to the cutter as possible, to provide a rigid tooling setup. Although arbor supports are not classified, a general rule of thumb can be used for arbor selection--the old reference type A is of a small bearing diameter, and the old reference type B is of a large bearing diameter. To prevent bending or springing of the arbor, you must install the arbor support before loosening or tightening the arbor nut.

(9) Size designation. All milling machines are identified by four basic factors: size, horsepower, model, and type. The size of a milling machine is based on the longitudinal (from left to right) table travel, in inches. Vertical, cross, and longitudinal travel are all closely related as far as the overall capacity. However, for size designation, only the longitudinal travel is used. There are six sizes of knee-type milling machines, with each number representing the number of inches of travel (See Table 8.1).

Table 8.1

Standard Size	Longitudinal Table Travel
No. 1	22 inches
No. 2	28 inches
No. 3	34 inches
No. 4	42 inches
No. 5	50 inches
No. 6	60 inches

If the milling machine in the shop is labeled No. 2HL, it has a table travel of 28 inches;

if it is labeled No. 5LD, it has a travel of 50 inches. The horsepower designation refers to the rating of the motor which is used to power the machine. The model designation is determined by the manufacturer and features vary with different brands. The type of milling machine is designated as plain or universal, horizontal or vertical, and knee and column, or bed. In addition, machines may have other special type designations and, therefore, may not fit any standard classification.

8.4 Milling Machine Accessories and Attachments

8.4.1 Arbors

Milling machine cutters can be mounted on several types of holding device. The machinist must know the devices, and the purpose of each to make the most suitable tooling setup for the operation to be performed. Technically, an arbor is a shaft on which a cutter is mounted. For convenience, since there are so few types of cutter holders that are not arbors, we will refer to all types of cutter holding devices as arbors.

(1) Description.

(a) Milling machine arbors are made in various lengths and in standard diameters of 7/8, 1, 1 1/4, and 1 1/2 inch. The shank is made to fit the tapered hole in the spindle, the other end is threaded.

NOTE: The threaded end may have left-handed or right-handed threads.

(b) Arbors are supplied with one of three tapers to fit the milling machine spindle, the milling machines Standard taper, the Brown and Sharpe taper, and the Brown and Sharpe taper with tang.

(c) The milling machine Standard taper is used on most machines of recent manufacture. It was originated and designed by the milling machine manufacturers to make removal of the arbor from the spindle much easier than will those of earlier design.

(d) The Brown and Sharpe taper is found mostly on older machines. Adapters or collets are used to adapt these tapers to fit the machines whose spindles have milling machine Standard tapers.

(e) The Brown and Sharpe taper with tang also is used on some of the older machines. The tang engages a slot in the spindle to assist in driving the arbor.

(2) Standard milling machine arbor.

(a) The Standard milling machine arbor has a straight, cylindrical shape, with a Standard milling taper on the driving end and a threaded portion on the opposite end to receive the arbor nut. One or more milling cutters may be placed on the straight cylindrical shaft of the arbor and held in position by means of sleeves and an arbor nut. The Standard milling machine arbor is usually splined and has keys, used to lock each cutter to the arbor shaft. Arbors are supplied in various lengths and standard diameters.

Chapter 8 Milling and Milling Machines

(b) The end of the arbor opposite the taper is supported by the arbor supports of the milling machine. One or more supports are used, depending on the length of the arbor and the degree of rigidity required. The end may be supported by alathe center, bearing against the arbor nut (Fig. 8.4 on the previous page) or by a bearing surface of the arbor fitting inside a bushing of the arbor support. Journal bearings are placed over the arbor in place of sleeves where an intermediate arbor support is positioned.

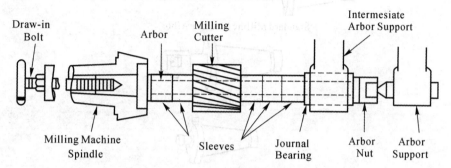

Fig. 8.4 Standard Milling Machine Arbor Installed

(c) The most common means of fastening the arbor in the milling machine spindle is by use of a draw-in bolt (Fig. 8.4). The bolt threads into the taper shank of the arbor to draw the taper into the spindle and hold it in place. Arbors secured in this manner are removed by backing out the draw-in bolt and tapping the end of the bolt to loosen the taper.

(3) Screw arbor (Fig. 8.5).

Screw arbors are used to hold small cutters that have threaded holes. These arbors have a taper next to the threaded portion to provide alignment and support for tools that require a nut to hold them against a tapered surface. A right-hand threaded arbor must be used for right-hand cutters; a left-hand threaded arbor is used to mount left-hand cutters.

(4) Slitting saw milling cutter arbor (Fig. 8.5).

The slitting saw milling cutter arbor is a short arbor having two flanges between which the milling cutter is secured by tightening a clamping nut. This arbor is used to hold the metal slitting saw milling cutters that are used for slotting, slitting, and sawing operations.

(5) End milling cutter arbor.

The end milling cutter arbor has a bore in the end in which the straight shank end milling cutters fit. The end milling cutters are locked in place by means of a setscrew.

(6) Shell end milling cutter arbor (Fig. 8.5).

Shell end milling arbors are used to hold and drive shell end milling cutters. The shell end milling cutter is fitted over the short boss on the arbor shaft and is held against the face of the arbor by a bolt, or a retaining screw. The two lugs on the arbor fit slots in the cutter to prevent the cutter from rotating on the arbor during the machining operation. A special wrench is used to tighten and loosen a retaining screw/bolt in the end of the arbor.

(7) Fly cutter arbor(Fig. 8.5).

The fly cutter arbor is used to support a single-edge lathe, shaper, or planer cutter bit, for boring and gear cutting operations on the milling machine. These cutters, which can be ground to any desired shape, are held in the arbor by a locknut. Fly cutter arbor shanks may have a Standard milling machine spindle taper, a Brown and Sharpe taper, or a Morse taper.

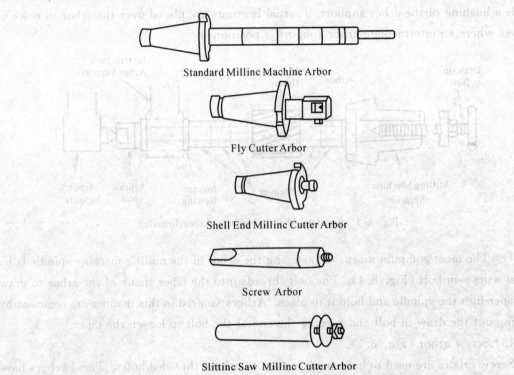

Fig. 8.5 Types of Milling Machne Arbors

8.4.2 Collets and Spindles

Milling cutters that contain their own straight or tapered shanks are mounted to the milling machine spindle with collets or spindle adapters which adapt the cutter shank to the spindle.

(1) Collets. Collets for milling machines serve to step up or increase the taper sizes so that small-shank tools can be fitted into large spindle recesses. They are similar to drilling machine sockets and sleeves except that their tapers are not alike. Spring collets are used to hold and drive straight-shanked tools. The spring collet chuck consists of a collet adapter, spring collets, and a cup nut. Spring collets are similar to lathe collets. The cup forces the collet into the mating taper, causing the collet to close on the straight shank of the tool. Collets are available in several fractional sizes.

(2) Spindle Adapters. Spindle adapters are used to adapt arbors and milling cutters to the standard tapers used for milling machine spindles. With the proper spindle adapters, any tapered or straight shank cutter or arbor can be fitted to any milling machine, if the sizes and tapers are standard.

Chapter 8 Milling and Milling Machines

8.4.3 Indexing Fixture (Fig. 8.6)

(1) It is an indispensable accessory for the milling machine, basically, a device for mounting workpieces and rotating them a specified amount around the workpiece's axis, as from one tooth space to another on a gear or cutter.

(2) The index fixture consists of an index head, also called a dividing head, and a footstock, similar to the tailstock of a lathe. The index head and the footstock are attached to the worktable of the milling machine by T-slot bolts. An index plate containing graduations is used to control the rotation of the index head spindle. The plate is fixed to the index head, and an index crank, connected to the index head spindle by a worm gear and shaft, is moved about the index plate. Workpieces are held between centers by the index head spindle and footstock. Workpieces may also be held in a chuck mounted to the index headspindle, or may be fitted directly into the taper spindle recess of some indexing fixtures.

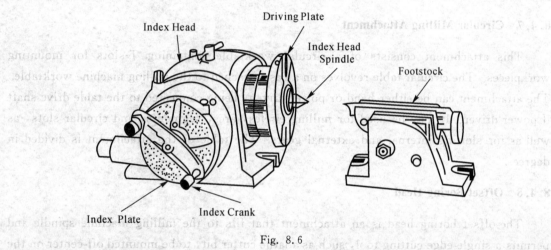

Fig. 8.6

(3) There are many variations of the indexing fixture. The name universal index head is applied to an index head designed to permit power drive of the spindle so that helixes may be cut on the milling machine.

"Gear cutting attachment" is another name for an indexing fixture; in this case, one primarily intended for cutting gears on the milling machine.

8.4.4 High-speed Milling Attachment

The rate of spindle speed of the milling machine may be increased from 1 1/2 to 6 times by the use of the high-speed milling attachment. This attachment is essential when using cutters and twist drills which must be driven at a high rate of speed in order to obtain an efficient surface speed. The attachment is clamped to the column of the machine and is driven by a set of gears from the milling machine spindle.

8.4.5 Vertical Spindle Attachment

This attachment converts the horizontal spindle of a horizontal milling machine to a vertical spindle. It is clamped to the column and driven from the horizontal spindle. It incorporates provisions for setting the bead at any angle, from the vertical to the horizontal, in a plane at right angles to the machine spindle. End milling and face milling operations are more easily accomplished with this attachment, due to the fact that the cutter and the surface being cut are in plain view.

8.4.6 Universal Milling Attachment

This device is similar to the vertical spindle attachment but is more versatile. The cutter head can be swiveled to any angle in any plane, whereas the vertical spindle attachment only rotates in one plane from the horizontal to the vertical.

8.4.7 Circular Milling Attachment

This attachment consists of a circular worktable containing T-slots for mounting workpieces. The circular table revolves on a base attached to the milling machine worktable. The attachment can be either hand or power driven, being connected to the table drive shaft if power driven. It may be used for milling circles, arcs, segments, and circular slots, as well as for slotting internal and external gears. The table of the attachment is divided in degrees.

8.4.8 Offset Boring Head

The offset boring head is an attachment that fits to the milling machine spindle and permits a single-edge cutting tool, such as a lathe cutter bit, to be mounted off-center on the milling machine. Workpieces can be mounted in a vise attached to the worktable and can be bored with this attachment.

8.5 Mounting and Indexing Work

8.5.1 General

(1) An efficient and positive method of holding workpieces to the milling machine table is essential if the machine tool is to be used to advantage. Regardless of the method used in holding, there are certain factors that should be observed in every case. The workpiece must not be sprung in clamping; it must be secured to prevent it from springing or moving away from the cutter; and it must be so aligned that it may be correctly machined.

(2) Milling machine worktables are provided with several T-slots, used either for clamping and locating the workpiece itself or for mounting various holding devices and

attachments. These T-slots extend the length of the table and are parallel to its line of travel. Most milling machine attachments, such as vises and index fixtures, have keys or tongues on the underside of their bases so that they may be located correctly in relation to the T-slots.

8.5.2 Methods of Mounting Workpieces

(1) Clamping a workpiece to the table. When clamping workpieces to the worktable of the milling machine, the table and workpiece should be free from dirt and burrs. Workpieces having smooth machined surfaces may be clamped directly to the table, provided the cutter does not come in contact with the table surface during the machining operation. When clamping workpieces with unfinished surfaces in this way, the table face should be protected by pieces of soft metal. Clamps should be placed squarely across the workpiece to give a full bearing surface. These clamps are held by Tslot bolts inserted in the T-slots of the table. Clamping bolts should be placed as near to the workpiece as possible. When it is necessary to place a clamp on an overhanging part of the workpiece, a support should be provided between the overhang and the table, to prevent springing or possible breakage. A stop should be placed at the end of the workpiece where it will receive the thrust of the cutter when heavy cuts are being taken.

(2) Clasping a workpiece to the angle plate. Workpieces clamped to the angle plate may be machined with surfaces parallel, perpendicular, or at an angle to a given surface. When using this method of holding a workpiece precautions should be taken, similar to those mentioned in (1) above for clamping the workpiece-directly to the table. Angle plates may be of either the adjustable or the nonadjustable type and are generally held in alignment by means of keys or tongues that fit into the table T-slots.

(3) Clamping workpieces in fixtures. Fixtures are generally used in production work where a number of identical pieces are to be machined. The design of the fixture is dependent upon the shape of the piece and the operations to be performed. Fixtures are always constructed to secure maximum clamping surfaces and are built to use a minimum number of clamps or bolts, in order to reduce the time required for setting up the workpiece. Fixtures should always be provided with keys to assure positive alignment with the table T-slots.

(4) Holding workpieces between centers. The indexing fixture is used to support workpieces which are centered on both ends. When the piece has been previously reamed or bored, it may be pressed upon a mandrel and then mounted between the centers, as with a lathe.

(a) There are two types of mandrels that may be used for mounting workpieces between centers. The solid mandrel is satisfactory for many operations, while the mandrel having a tapered shank is preferred when fitting the workpiece into the indexing head of the spindle.

(b) A jack screw is used to prevent springing of long slender workpieces held between centers, or workpieces that extend some distance from the chuck.

(c) Workpieces mounted between centers are fixed to the index head spindle by means of a lathe dog. The bent tail of the dog should be fastened between the setscrews provided in the driving center clamp in such a manner as to avoid backlash and prevent springing the mandrel. When milling certain types of workpieces a milling machine dog may be used to advantage. The tail of the dog is held in a flexible ball joint which eliminates springing or shaking of the workpiece and/or the dog. The flexible ball joint allows the tail of the dog to move in a radius along the axis of the workpiece, making it particularly useful in the rapid milling of tapers.

(5) Holding workpieces in a chuck. Before screwing the chuck to the index head spindle, it should be cleaned and all burrs removed from the spindle or the chuck. Burrs may be removed with a smooth cut, three-cornered file or scraper. Cleaning should be accomplished with a piece of spring-steel wire bent and formed to fit the angle of the threads, or by the use of compressed air. The chuck should not be tightened on the spindle so tightly that a wrench or bar is required to remove it. Cylindrical workpieces, held in the universal chuck, may be checked for trueness by using a test indicator mounted on a base which rests on the milling machine.

The indicator point should contact the circumference of small diameter workpieces, or the circumference and exposed face of large diameter pieces. While checking for trueness, the workpiece should be revolved by rotating the index head spindle.

(6) Holding workpieces in the vise. Three types of vises are manufactured in various sizes for holding milling machine workpieces. These vises have locating keys or tongues on the underside of their bases so they may be located correctly in relation to the T-slots on the milling machine table.

(a) The plain vise, similar to the machine table vise, is used for milling straight workpieces; it is bolted to the milling machine table at right angles or parallel to the machine arbor.

(b) The swivel vise can be rotated and contains a scale graduated in degrees at its base to facilitate milling workpieces at any angle on a horizontal plane. This vise is fitted into a graduated circular base fastened to the milling machine table and located by means of keys placed in the T-slots. By loosening the bolts, which clamp the vise to its graduated base, the vise may be moved to hold the workpiece at any angle in a horizontal plane. To set a swivel vise accurately with the machine spindle, a test indicator should be clamped to the machine arbor and a check made to determine the setting by moving either the transverse or the longitudinal feeds, depending upon the position of the vise jaws. Any deviation as shown by the test indicator should be corrected by swiveling the vise on its base.

(c) The universal vise is constructed to allow it to be set at any angle, either horizontally or vertically, to the axis of the milling machine spindle. Due to the flexibility of this vise, it is not adaptable for heavy milling (Fig. 8.7).

(d) When rough or unfinished workpieces are to be vise mounted, a piece of protecting

Chapter 8 Milling and Milling Machines

material should be placed between the vise jaws and the workpiece to eliminate marring the jaws.

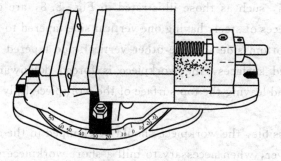

Fig. 8.7 Universal Vise

(e) When it is necessary to position a workpiece above the vise jaws, parallels of the same size and of the proper height should be used (Fig. 8.8). These parallels should only be high enough to allow the required cut, as excessive raising reduces the holding ability of the jaws. When holding a workpiece on parallels, a soft lead hammer should be used to tap the top surface of the piece after the vise jaws have been tightened. This tapping should be continued until the parallels cannot be moved by hand. After once set, additional tightening has a tendency to raise the work off the parallels.

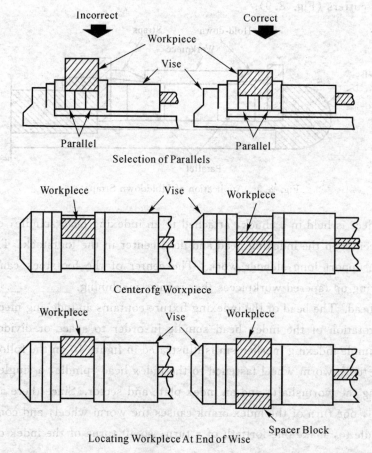

Fig. 8.8 Mounting Workpiece in the Vise

— 125 —

(f) If the workpiece is so thin that it is impossible to let it extend over the top of the vise, hold-down straps, such as those illustrated in Fig. 8.8, are generally used. These straps are hardened pieces of steel, having one vertical side tapered to form an angle of about 92 degrees with the bottom side and the other vertical side tapered to a narrow edge. By means of these tapered surfaces, the workpiece is forced downward onto the parallels, holding them firmly and leaving the top surface of the workpiece fully exposed to the milling cutter.

(g) Whenever possible, the workpiece should be clamped in the center of the vise jaws (see Fig. 8.8); however, when necessary to mill a short workpiece which must be held at the end of the vise, a spacing block of the same thickness as the piece (see Fig. 8.8) should be placed at the opposite ends of the jaws. This will avoid strain on the movable jaw and prevent the piece from slipping.

8.5.3 Indexing the Workpieces.

(1) General. Indexing equipment is used to hold the workpiece, and to provide a means of turning it so that a number of accurately located speed cuts can be made, such as those required in cutting tooth spaces on gears, milling grooves in reamers and taps, and forming teeth on milling cutters (Fig. 8.9).

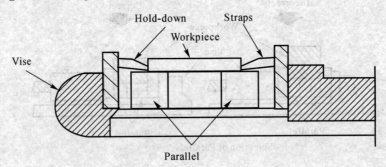

Fig. 8.9 Application of Holddown Straps

The workpiece is held in a chuck, attached to an indexing head spindle, or mounted in between a live center in the indexing head and dead center in the footstock. The center rest can be used to support long slender work. The center of the footstock can be raised or lowered for setting up tapered workpieces that require machining.

(2) Index head. The bead of the indexing fixture contains an indexing mechanism, used to control the rotation of the index head spindle in order to space or divide a workpiece accurately. A simple indexing mechanism is illustrated in figure 10 on the following page. It consists of a 40-tooth worm wheel fastened to the index head spindle, a single-cut worm, a crank for turning the wormshaft, and an index plate and sector. Since there are 40 teeth in the worm wheel, one turn of the index crank causes the worm wheel, and consequently the index head spindle to, make one-fortieth of a turn; so 40 turns of the index crank revolves

the spindle one full turn (Fig. 8.10).

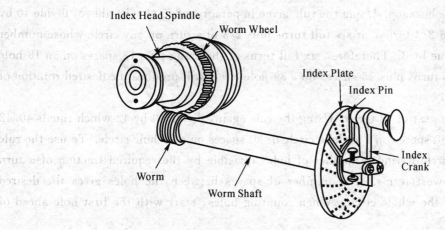

Fig. 8.10 Simple Indexing Mechanism

(3) Plain indexing. The following principles apply to basic indexing of workpieces:

(a) Suppose it is desired to mill a spur gear with 8 equally spaced teeth. Since 40 turns of the index crank will turn the spindle one full turn, one-eighth of 40, or 5 turns of the crank after each cut, will space the gear for 8 teeth.

(b) The same principle applies whether or not the divisions required divide evenly into 40. For example, if it is desired to index for 6 divisions, 6 divided into 40 equals 6 2/3 turns; similarly, to index for 14 spaces, 14 divided into 40 equals 2 6/7 turns. Therefore, the following rule can be derived: to determine the number of turns of the index crank needed to obtain one division of any number of equal divisions on the workpiece, divide 40 by the number of equal divisions desired (provided the worm wheel has 40 teeth, which is standard practice).

(4) Index plate. The index plate (Fig. 8.11) is a round metal plate with a series of six or more circles of equally spaced holes; the index pin on the crank can be inserted in any hole in any circle.

With the interchangeable plates regularly furnished with most index heads, the spacings necessary for most gears, boltheads, milling cutters, splines, and so forth, can be obtained. The following sets of plates are standard equipment:

Brown and sharpe type, 3 plates of 6 circles, each drilled as follows:
Plate 1- 15, 16, 17, 18, 19, 20 holes.
Plate 2- 21, 23, 27, 29, 31, 33 holes.
Plate 3- 37, 39, 41, 43, 47, 49 holes.
Cincinnati type, one plate drilled on both sides with circles divided as follows:
First side- 24, 25, 28, 30, 34, 37, 38, 39, 41, 42, 43 holes.
Second side- 46, 47, 49, 51, 53, 54, 57, 58, 59, 62, 66 holes.

(5) Indexing operation. The two following examples show how the index plate is used

to obtain any desired part of a whole spindle turn by plain indexing.

(a) To mill a hexagon. Using the rule given in paragraph 8.5 (3)(b) above, divide 40 by 6, which equals 6 2/3 turns, or six full turns plus 2/3 of a turn on any circle whose number of holes is divisible by 3. Therefore, six full turns of the crank plus 12 spaces on an 18-hole circle, or six full turns plus 26 spaces on a 39-hole circle will produce the desired rotation of the workpiece.

(b) To cut a gear of 42 teeth. Using the rule again, divide 40 by 42 which equals 40/42 or 20/21 turns, 40 spaces on a 42-hole circle or 20 spaces on a 21-hole circle. To use the rule given, select a circle having a number of holes divisible by the required fraction of a turn reduced to its lowest terms. The number of spaces between the holes gives the desired fractional part of the whole circle. When counting holes, start with the first hole ahead of the index pin.

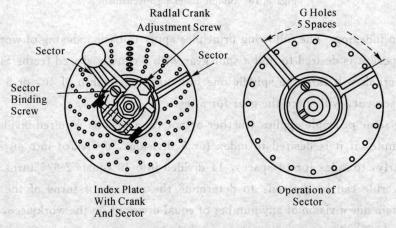

Fig. 8.11 Index Plate and Sector

(6) Sector. The sector (Fig. 8.11) indicates the next hole in which the pin is to be inserted and makes it unnecessary to count the holes when moving the index crank after each cut. It consists of two radial, beveled arms which can be set at any angle to each other and then moved together around the center of the index plate. Assume that it is desired to make a series of cuts, moving the index crank 1 1/4 turns after each cut. Since the circle has 20 turns, the crank must be turned one full turn plus 5 spaces after each cut. Set the sector arms to include the desired fractional part of a turn, or 5 spaces, between the beveled edges of its arms. If the first cut is taken with the index pin against the left-hand arm, to take the next cut, move the pin once around the circle and into the hole against the right-hand arm of the sector. Before taking the second cut, move the arms so that the left-hand arm is again against the pin; this moves the right-hand arm another five spaces ahead of the pin. Then take the second cut; repeat the operation until all the cuts have been completed.

NOTE: It is a good practice always to index clockwise on the plate.

(7) Direct indexing. The construction of some index heads permits the worm to be

disengaged from the worm wheel, making possible a quicker method of indexing, called direct indexing. The index head is provided with a knob which, when turned through part of a revolution, operates an eccentric and disengages the worm. Direct indexing is accomplished by an additional index plate fastened to the index head spindle. A stationary plunger in the index head fits the holes in the index plate. By moving the plate by hand to index directly, the spindle and the workpiece rotate an equal distance.

Direct index plates usually have 24 holes and offer a quick means of milling squares, hexagons, taps, etc. Any number of divisions which is a factor of 24 can be indexed quickly and conveniently by the direct indexing method.

(8) Differential indexing. Sometimes a number of divisions are required which cannot be obtained by simple indexing with the index plates regularly supplied. To obtain these divisions a differential index head is used. The index crank is connected to the wormshaft by a train of gears instead of by a direct coupling and with simple indexing. The selection of these gears involves calculations similar to those used in calculating change gear ratio for cutting threads on a lathe.

(9) Angular Indexing.

(a) When you must divide work into degrees or fractions of degrees by plain indexing, remember that one turn of the index crank will rotate a point on the circumference of the work 1/40 of a revolution. Since there are 360° in a circle, one turn of the index crank will revolve the circumference of the work 1/40 of 360°, or 9°. Hence, in using the index plate and fractional parts of a turn, 2 holes in a 18-hole circle equals 10, 1 hole in a 27-hole circle equals 2/3°, 3 holes in a 54-hole circle equals 1/3°. To determine the number of turns, and parts of a turn of the index crank for a desired number of degrees, divide the number of degrees by 9.

The quotient will represent the number of complete turns and fractions of a turn that you should rotate the index crank. For example, the calculation for determining 15° when an index plate with a 54-hole circle is available, is as follows:

$$\frac{15}{9} = 1 + \frac{6}{9} \times \frac{6}{6} = 1\frac{36}{54}$$

or one complete turn plus 36 holes on the 54-hole circle. The calculation for determining 13 1/2° when an index plate with an 18-hole circle is available, is as follows:

$$\frac{13.5}{9} = 1\frac{4.5}{9} \times \frac{2}{2} = 1\frac{9}{18}$$

(b) When indexing angles are given in minutes and approximate divisions are acceptable, movement of the index crank and the proper index plate may be determined by the following calculations:

You can determine the number of minutes represented by one turn of the index crank by multiplying the number of degrees covered in one turn of the index crank by 60:

$$9° \times 60 = 540°$$

Therefore, one turn of the index crank will rotate the index head spindle 540 minutes.

(c) The number of minutes (540) divided by the number of minutes in the division desired, indicates the total number of holes required in the index plate used. (Moving the index crank one hole will rotate the index spindle through the desired number of minutes of the angle.) This method of indexing can be used only for approximate angles since ordinarily the quotient will come out in mixed numbers, or in numbers for which no index plate is available. However, when the quotient is nearly equal to the number of holes in an available index plate, the nearest number of holes can be used and the error will be very small. For example, the calculation for 24 minutes would be:

$$\frac{540}{24} = \frac{22.5}{1}$$

or one hole on the 22.5 - hole circle. Since there is no 22.5 - hole circle on the index plate, a 23-hole circle plate would be used.

(d) If a quotient is not approximately equal to an available circle of holes, multiply by any trial number which will give a product equal to the number of holes in one of the available index circles. You can then move the crank the required number of holes to give the desired division. For example, the calculation for determining 54 minutes when an index plate that has a 20-hole circle is available, is as follows:

$$\frac{54}{540} = \frac{1}{10} \times \frac{2}{2} = \frac{2}{20} \frac{\text{(no. of holes)}}{\text{(20 - hole circle)}}$$

or 2 holes on the 20-hole circle.

8.6 Milling Machine Operations

1. General

The milling machine is one of the most versatile metalworking machines in a shop. It is capable of performing simple operations, such as milling a flat surface or drilling a hole, or more complex operations, such as milling helical gear teeth. It would be impractical to attempt to discuss all of the operations that a milling machine can do. The success of any milling operation depends to a great extent upon judgment in setting up the job, selecting the proper cutter, and holding the cutter by the best means. Even though we will discuss only the more common operations, the machinist will find that by using a combination of operations, he will be able to produce a variety of work projects. Some fundamental practices have been proved by experience to be necessary for good results on all jobs. Some of these practices are mentioned below.

(1) Before setting up a job, be sure that the workpiece, the table, the taper in the spindle, and arbor or cutter shank, are all clean and free from chips, nicks, or burrs.

(2) Set up every job as close to the milling machine spindle as the circumstances permit.

(3) Do not select a milling cutter of larger diameter than is necessary.

(4) Keep milling cutters sharp at all times.

(5) Do not change feeds or speeds while the milling machine is in operation.

(6) Always lower the table before backing the workpiece under a revolving milling cutter.

(7) Feed the workpiece in a direction opposite to the rotation of the milling cutter, except when milling long or deep slots or when cutting off stock.

(8) Never run a milling cutter backwards.

(9) When using clamps to secure the workpieces, be sure that they are tight and that the workpiece is held so that it will not spring or vibrate while it is being cut.

(10) Use a recommended cutting oil liberally.

(11) Keep chips away from the workpiece; brush them out of the way by any convenient means, but do not do so by hand or with waste.

(12) Use good judgment and common sense in planning every job, and profit by previous mistakes.

2. Operations

Milling operations may be classified under four general headings as follows:

(1) Face Milling - machining flat surfaces which are at right angles to the axis of the cutter.

(2) Plain or Slab Milling - machining flat surfaces which are parallel to the axis of the cutter.

(3) Angular Milling - machining flat surfaces which are at an inclination to the axis of the cutter.

(4) Form Milling - machining surfaces having an irregular outline.

3. Speeds for Milling Cutters

(1) General. The speed of a milling cutter is the distance in feet per minute that each tooth travels as it cuts its chips. The number of spindle revolutions per minute necessary to give a desired peripheral speed on the size of the milling cutter. The best speed is determined by the type of material being cut and the size and type of cutter used. The smoothness of the finish desired and the power available are other factors relating to the cutter speed.

(2) Selecting proper cutting speed.

(a) The approximate values given in Table 1 on the following page may be used as a guide for selecting the proper cutting speed. If the operator finds that the machine, the milling cutter, or the workpiece cannot be handled suitably at these speeds, immediate readjustment should be made.

(b) Table 8.2 lists speeds for high-speed steel milling cutters. If carbon steel cutters are used, the speed should be about one-half the speed recommended in the table. If carbide-tipped cutters are used, the speed can be doubled.

(c) If a plentiful supply of cutting oil is applied to the milling cutter and the workpiece, the speeds can be increased from 50 to 100 percent.

MILLING MACHINE CUTTING SPEEDS FOR HIGH-SPEED STEEL MILLING CUTTERS.

Table 8.2

Materal	Cutting SPeed			
	Plain milling cutters		End milling cutters	
	Roughing	Finishing	Roughing	Finishing
Aluminum···	400 to 1,000	400 to 1,000	400 to 1,000	400 to 1,000
Brass, composition···	125 to 200	90 to 200	90 to 150	90 to 150
Brass, yellow···	150 to 200	100 to 250	100 to 200	100 to 200
Bronze, phosphor and	30 to 80	25 to 100	30 to 80	30 to 80
Cast iron (hard)···	25 to 40	10 to 30	25 to 40	20 to 45
Cast iron (soft and mdeium)	40 to 75	25 to 80	35 to 65	30 to 80
Monel metal···	50 to 75	50 to 75	40 to 60	40 to 60
Steel, bard···	25 to 50	25 to 70	25 to 50	25 to 70
Steel, soft···	60 to 120	45 to 110	50 to 85	45 to 100

(d) For roughing cuts, a moderate speed and coarse feed give best results; for finishing cuts, the best practice is to reverse these conditions, using a higher speed and a lighter cut.

(3) Speed Computation.

(a) The formula for calculating spindle speed in revolutions per minute is as follows:

$$\text{rpm} = \frac{cs \times 4}{D}$$

Where, rpm = spindle speed (in revolution per minute),

cs = cutting speed of milling cutter (in surface feet per minute),

D = diameter of milling cutter (in inches).

For example, the spindle speed for machining a piece of steel at a speed of 35 rpm with a cutter 2 inches in diameter is calculated as follows:

$$\text{rpm} = \frac{cs \times 4}{D} = \frac{35 \times 4}{2} = 70$$

Therefore, the milling machine spindle would be set for as near 70 rpm as possible. If the calculated rpm cannot be obtained, the next lower selection should be made.

(b) Table 8.3 is provided to facilitate spindle speed computations for standard cutting speeds and standard milling cutters.

Table 8.3 Milling Cutter Rotatinal Speeds

Diameter of Cutter	Cutter Speed													
	25	30	35	40	50	60	70	80	90	100	120	140	180	200
	Cutter revolutionper minute													
1/16	382	458	535	611	764	917	1,070	1,222	1,376	1,528	1,834	2,139	2,445	3,056
1/8	306	357	428	489	611	733	856	978	1,100	1,222	1,466	1,711	1,055	2,444
2/8	255	306	357	408	309	611	713	815	916	1,018	1,222	1,425	1,629	2,036
3/8	218	262	306	349	437	524	611	699	786	874	1,049	1,224	1,398	1,748
1/2	191	229	268	306	382	459	535	611	688	764	917	1,070	1,222	1,528
5/8	153	184	214	245	306	367	428	480	552	612	736	857	979	1,224
6/8	127	153	178	203	254	306	357	408	458	508	610	711	813	1,016
7/8	109	131	153	175	219	282	306	349	392	438	526	613	701	876
1	95.5	115	134	153	191	229	267	306	344	382	458	535	611	764
1 1/4	76.3	91.8	107	123	153	183	214	245	274	306	367	428	490	612
1 1/2	63.7	76.3	89.2	102	127	153	176	204	230	254	305	358	406	508
1 3/4	54.5	65.5	76.4	87.3	109	131	153	175	196	218	282	305	349	438
2	47.8	57.3	66.9	76.4	95.5	115	134	153	172	191	229	267	306	382
2/2	38.2	45.8	53.5	61.2	76.3	91.7	107	122	138	133	184	213	245	306
3	31.8	38.2	44.6	51	63.7	76.4	89.1	102	114	127	152	178	203	254
3 1/2	27.3	32.7	38.2	43.6	54.5	85.5	76.4	87.4	98.1	109	131	153	174	218
4	23.9	28.7	33.4	38.2	47.8	57.3	66.9	76.4	86	95.6	115	134	153	191
5	19.1	22.9	26.7	30.6	38.2	45.9	53.6	61.1	68.8	76.4	91.7	107	122	153
6	15.9	19.1	22.3	25.5	31.8	38.2	44.6	51.0	57.2	63.6	76.3	89	102	127
7	13.6	16.4	19.1	21.8	27.3	32.7	38.2	43.7	49.1	54.6	65.5	76.4	87.4	109
8	11.9	14.3	16.7	19.1	23.9	28.7	33.4	38.2	43	47.8	67.4	66.9	76.5	95.6

4. Feeds For Milling

(1) General.

The rate of feed, or the speed at which the workpiece passes the cutter, determines the time required for cutting a job. In selecting the feed, there are several factors which should be considered. These factors are:

(a) Forces are exerted against the workpiece, the cutter, and their holding devices during the cutting process. The force exerted varies directly with the amount of metal removed and can be regulated by the feed and the depth of cut.

Therefore, the correct amount of feed and depth of cut are interrelated, and in turn are dependent upon the rigidity and power of the machine. Milling machines are limited by the power that they can develop to turn the cutter and the amount of vibration they can resist when using coarse feeds and deep cuts.

(b) The feed and depth of cut also depend upon the type of milling cutter being used.

For example, deep cuts or coarse feeds should not be attempted when using a small diameter end milling cutter, as such an attempt would spring or break the cutter. Coarse cutters with strong cutting teeth can be fed at a faster rate because the chips may be washed out more easily by the cutting oil.

(c) Coarse feeds and deep cuts should not be used on a frail workpiece, or on a piece that is mounted in such a way that its holding device is not able to prevent springing or bending.

(d) The degree of finish required often determines the amount of feed. Using a coarse feed, the metal is removed more rapidly but the appearance and accuracy of the surface produced may not reach the standard desired for the finished product. Because of this, finer feeds and increased speeds are used for finer, more accurate finishes. Most mistakes are made through overspeeding, underspeeding, and overfeeding. Overspeeding may be detected by the occurrence of a squeaking, scraping sound. If vibration (referred to as "chattering") occurs in the milling machine during the cutting process, the speed should be reduced and the feed increased. Too much cutter clearance, a poorly supported workpiece, or a badly worn machine gear are common causes of "chattering."

(2) Typical Feeds.

(a) Feed for milling cutters will generally run from 0.002 to 0.250 inch per cutter revolution, depending upon the diameter of the cutter, the kind of material, the width and depth of the cut, the size of the workpiece, and the condition of the machine.

(b) Good finishes may be obtained using a 3-inch plain milling cutter at a 40 feet per minute speed, with a feed of 0.040-inch per cutter revolution.

(3) Direction of Feed.

(a) It is usually regarded as standard practice to feed the workpiece against the milling cutter (Fig. 8.12 on the following page). When the piece is fed against the milling cutter, the teeth cut under any scale on the workpiece surface and any backlash in the feed screw is taken up by the force of cut.

(b) As an exception to this recommendation, it is advisable to feed with the milling cutter (Fig. 8.12), when cutting off stock, or when milling comparatively deep or long slots.

(c) The direction of cutter rotation is related to the manner in which the workpiece is held. The cutter should rotate so that the piece springs away from the cutter; then there will be no tendency for the force of the cut to loosen the workpiece. No milling cutter should be rotated backward as this will break the teeth. If it is necessary to stop the machine during a finishing cut, the power feed should never be thrown out, nor should the workpiece be feedback under the cutter, unless the cutter is stopped or the workpiece lowered. Never change feeds while the cutter is rotating.

Chapter 8 Milling and Milling Machines

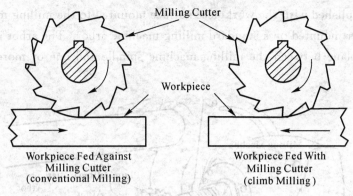

Fig. 8.12 Direction of Feed for Milling

5. Cutting Oils

(1) The major advantage of a cutting oil is that it reduces frictional heat, thereby giving longer life to the cutting edges of the teeth. The oil also serves to lubricate the cutter face and to flush away the chips, consequently reducing the possibility of marring the finish.

(2) Cutting oil compounds for various metals are given in tablebelow. In general, a simple coolant is all that is required for roughing. Finishing requires a cutting oil with a good lubricating properties to help produce a good finish on the workpiece. Aluminum and cast iron are almost always machined dry.

(3) The cutting oil or coolant should be directed, by means of a coolant drip can or a pump system, to the point where the cutter contacts the workpiece. Regardless of the method used, the cutting oil should be allowed to flow freely over the workpiece and the cutter (See Table 8.4).

Table 8.4

Material	Cutting fluid	
	Roughing	Finishing
Aluminum		Dry
Brass, Composition	Dry	Dry; turpentine
Brass, yellow	Dry; soda water mixture	Dry
Bronze, phosphor and manganese	Dry; soda water mixture Soluble cutting oil	Suffurized fatty mineral cutting oil; pure lard cutting oil.
Cast iron (hard)	Dry; soda water mixture	Dry; turrpentine
Cast iron (soft & medium)	Dry	Dry
Monel metal	Dry; soluble cutting oil; mineral fatty blend cutting oil	Mineral fatty blend cutting oil
Steel, hard	Soluble cutting oil	Sulfurized fatty mineral cutting oil; pure lard cutting oil
Steel, soft		Mineral fatty blend cutting oil

6. Plain Milling

(1) General. Plain milling, also called surface milling and slab milling, is milling flat surfaces with the milling cutter axis parallel to the surface being milled. Generally, plain

milling is accomplished with the workpiece surface mounted to the milling machine table and the milling cutter mounted on a standard milling machine arbor. The arbor is well supported in a horizontal plane between the milling machine spindle and one or more arbor supports (Fig. 8.13).

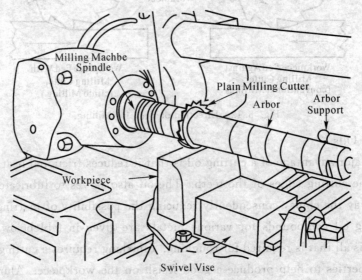

Fig. 8.13　Plain Milling Operations

(2) Operation. For plain milling, the workpiece is generally clamped directly to the table or supported in a vise. The milling machine table should be checked for alignment before starting to make a cut. If the workpiece surface that is to be milled is at an angle to the base plane of the piece, the workpiece should be mounted in a universal vise or an adjustable angle plate. The holding device should be adjusted so that the workpiece surface is parallel to the table of the milling machine.

(a) For plain milling operations, a plain milling cutter should be used. Deeper cuts may generally be taken when narrow cutters are used than with wide cutters. The choice of milling cutters should be based on the size of the workpiece. If a wide area is to be milled, fewer traverses will be required using a wide cut.

(b) A typical setup for plain milling is illustrated in figure 13 on page 37. Note, the milling cutter is positioned on the arbor with sleeves so that it is as close as possible to the milling machine spindle, while maintaining sufficient clearance between the vise and the milling machine column. This practice reduces torque in the arbor and permits more rigid support for the cutter.

(c) If large quantities of metal are to be removed, a coarse tooth cutter should be used for roughing and a finer tooth cutter should be used for finishing. A relatively slow cutting speed and a fast table feed should be used for roughing, and a relatively fast cutting speed, and a slow table feed used for finishing. The surface should be checked for accuracy after each completed cut.

7. Angular Milling

(1) General.

Angular milling, or angle milling, is milling flat surfaces which are neither parallel nor perpendicular to the axis of the milling cutter. A single-angle milling cutter (Fig. 8.14) is used for this operation. Milling dovetails is a typical example of angular milling. When milling dovetails, the usual angle of the cutter is 45°, 50°, 55°, or 60°, based on common dovetail designs.

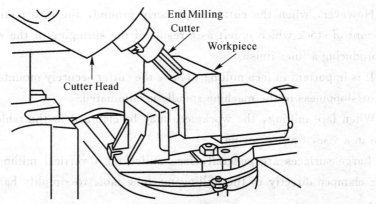

Fig. 8.14 Angular Face Milling

(2) Operation.

(a) When cutting dovetails on the milling machine, the workpiece may be held in the vice, clamped to the table, or clamped to an angle plate. Fig. 8.15 shows the workpiece mounted to a lathe faceplate for angular milling with the milling and grinding lathe attachment. The tongue or groove is first roughed-out using a side milling cutter, after which the angular sides and base are finished with an angle cutter.

(b) In general practice, the dovetail is laid out on the workpiece surface before the milling operation is started. To do this, the required outline should be inscribed and the line prick punched. These lines and punch marks may then be used as a guide during the cutting operation.

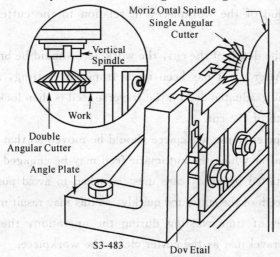

Fig. 8.15 Angular Milling

8. Face Milling

(1) General. Face milling, also called end milling and side milling, is machining surfaces perpendicular to the axis of the cutter.

(2) Operation. Face milling cutters, end milling cutters (Fig. 8.16), and side milling cutters are used for face milling operations. The size and nature of the workpiece determines the type and size of the cutter required.

(a) In face milling, the teeth on the periphery of the cutter do practically all of the cutting. However, when the cutter is properly ground, the face teeth actually remove a small amount of stock which is left as a result of the springing of the workpiece or cutter, thereby producing a finer finish.

(b) It is important in face milling to have the cutter securely mounted and to see that all end play or sloppiness in the machine spindle is eliminated.

(c) When face milling, the workpiece may be clamped to the table or angle plate, or supported in a vise, fixture, or jig.

(d) Large surfaces are generally face milled on a vertical milling machine with the workpiece clamped directly to the milling machine table to simplify handling and clamping operations.

(e) Fig. 8.16 illustrates face milling performed with a swivel cutter head milling machine with its spindle in a vertical position. The workpiece is supported parallel to the table in a swivel vise.

(f) Angular surfaces can also be face milled on a swivel cutter head milling machine. In this case, the workpiece is mounted to the table and the cutter head is swiveled to bring the end milling cutter perpendicular to the surface to be produced.

(g) During face milling operations, the workpiece should be fed against the milling cutter so that the pressure of the cut is downward, thereby holding the work against the table.

(h) Whenever possible, the edge of the workpiece should be in line with the center of the cutter. This position of the workpiece, in relation to the cutter, will help eliminate slippage.

(i) When setting the depth of the cut, the workpiece should be brought up to just touch the revolving cutter. After a cut has been made from this setting, a measurement of the workpiece is taken. The graduated dial on the traverse feed is then locked and used as a guide in determining the depth of the cut.

①When starting the cut, the workpiece should be moved so that the cutter is nearly in contact with its edge, after which the automatic feed may be engaged.

②When a cut is started by hand, care must be taken to avoid pushing the corner of the workpiece between the teeth of the cutter too quickly, as this may result in cutter tooth breakage.

③In order to prevent time wasting during the operation, the feed trips should be adjusted to stop table travel just as the cutter clears the workpiece.

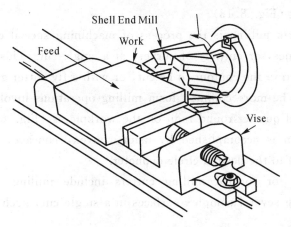

Fig. 8.16 Face Milling

9. Straddle Milling

(1) General. When two or more parallel vertical surfaces are machined at a single cut, the operation is called straddle milling. Straddle milling is accomplished by mounting two side milling cutters on the same arbor, set apart so that they straddle the workpiece.

(2) Operation. Fig. 8.17 illustrates a typical example of straddle milling. In this case a spline is being cut, but the same operation may be applied when cutting squares or hexagons on the end of a cylindrical workpiece. The workpiece is usually mounted between centers in the indexing fixture, or mounted vertically in a swivel vise. The two side milling cutters are separated by spacers, washers, and shims so that the distance between the cutting teeth of the cutters is exactly equal to the width of the workpiece area required. When cutting a square by this method, two opposite sides of the square are cut, then the spindle of the indexing fixture or the swivel vise is rotated 90° and the other two sides of the workpiece are straddle milled.

10. Gang Milling (Fig. 8.17)

Gang milling is the term applied to an operation in which two or more milling cutters are used together on one arbor when cutting horizontal surfaces. The usual method is to mount two or more milling cutters of different diameters, shapes and/or widths on an arbor as shown in the following page. The possible cutter combinations are unlimited and are determined in each case by the nature of the job.

Fig. 8.17

11. Form Milling (Fig. 8.18)

(1) General. Form milling is the process of machining special contours composed of curves and straight lines, or entirely of curves, at a single cut. This is done with formed milling cutters, shaped to the contour to be cut, or with a fly cutter ground for the job.

(2) Operation. The more common form milling operations involve milling half-round recesses and beads and quarter-round radii on the workpieces (Fig. 8.19 on the following page). This operation is accomplished by using convex, concave, and corner rounding milling cutters ground to the desired circle diameter.

(3) Other jobs for formed milling cutters include milling intricate patterns on workpieces and milling several complex surfaces in a single cut, such as produced by gang milling.

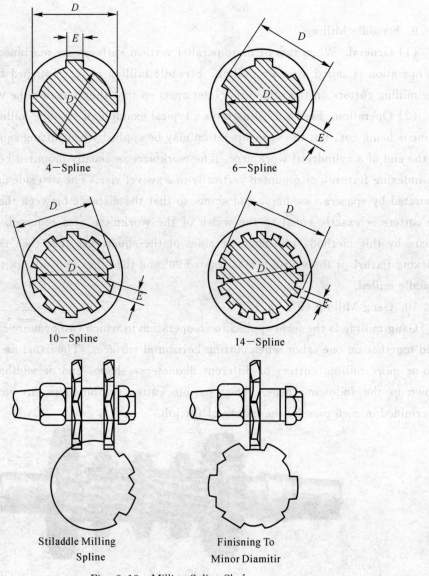

Fig. 8.18 Milling Spline Shafts

Woodruff Keyway Milling (Fig. 8.19).

(1) General.

Keyways are machined grooves of different shapes, cut along the axis of the cylindrical surface of shafts, into which keys are fitted to provide a positive method of locating and driving members mounted on the shafts. A keyway is also machined on the mounted member to receive the key.

The type of key and corresponding keyway to be used depends on the class of work for which it is intended. The most commonly used type of key is the woodruff.

(2) Operation.

(a) Woodruff keys are semi-cylindrical in shape and are manufactured in various diameters and widths. The circular side of the key is seated into a keyway which is milled into a shaft with a woodruff keyslot milling cutter having the same diameter as that of the key.

(b) Woodruff key sizes are designated by a code number in which the last two digits indicate the diameter of the key in eighths of an inch. These digits precede the last two digits and give the width of the key in thirty seconds of an inch. Thus, a number 204 woodruff key would be 4/8 or 1/2 inch in diameter and 2/32 or 1/16 inch wide; a number 1012 woodruff key would be 12/8 or 1 1/2 inches in diameter and 10/32 or 5/16 inch wide.

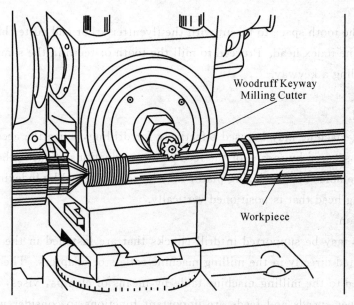

Fig. 8.19   Woodruff Keyway Milling

(3) Milling Woodruff Keyslots.

The woodruff keyslot milling cutter is mounted in a spring collet or adapter which has been inserted in the spindle of the milling machine or milling attachment (Fig. 8.21). With the milling cutter located over the position in which the keyway is to be cut, the workpiece should be raised, or the cutter lowered, until the peripheral teeth come in contact with the workpiece. At this point the graduated dial on the vertical feed adjustment should be locked and the clamp on the table set. Using the vertical feed, with the graduated dial as a guide, the workpiece is raised or the cutter lowered, until the teeth come in contact with the workpiece. At this point, the graduated dial on the vertical feed adjustment should be locked and the clamp on the table set. Using the vertical feed, and with the graduated dial as a guide, the workpiece is raised, or the cutter lowered, until the full depth of the keyslot is cut, completing the operation.

NOTE

This method of gear cutting is not as accurate as using an involute gear cutter and should be used only for emergency cutting of teeth.

(a) Fasten the indexing fixture to the milling machine table. Use a mandrel to mount the gear between the index head and the footstock centers. Adjust the indexing fixture on the milling machine table, or adjust the position of the cutter, to make the gear axis perpendicular to the milling machine spindle axis.

(b) Take the cutter bit that has been ground to the shape of the gear tooth spaces and fasten it in the flycutter arbor. Adjust the cutter centrally with the axis of the gear. Rotate the milling machine spindle to position the cutter bit in the flycutter so that its cutting edge is downward.

(c) Align the tooth space to be cut with the flycutter arbor and cutter bit by turning the index crank on the index head. Proceed to mill the tooth or teeth in the same manner as you would when milling a keyway.

12. Drilling

(1) General.

The milling machine may be used effectively for drilling, since the accurate location of the hole may be secured by means of the feed screw graduations. Spacing holes in a circular path, such as the holes in an indexing plate, may be accomplished by indexing the workpiece with the indexing head that is positioned vertically.

(2) Operation.

Twist drills may be supported in drill chucks that are fastened in the milling machine spindle or mounted directly in the milling machine collets or adapters. The workpiece to be drilled is fastened to the milling machine table by means of clamps, vises, or angle plates. Remember, proper speeds and feeds are important functions to consider when performing drilling operations on the milling machine.

13. Boring

Various types of boring toolholders may be used for boring on the milling machine. The boring tool can either be a straight shank, held in chucks and holders, or tapered shanks to fit collets and adapters. The two attachments most commonly used for boring are the flycutter arbor and the offset boring head. The single-edge cutting tool that is used for boring on the milling machine is the same as a lathe cutter bit. Cutting speeds, feeds, and depth of cut should be the same as those prescribed for lathe operations (See Table 8.5 and Table 8.6).

Table 8.5 INVOLUTE GEAR MILLING CUTTERS

Number of Cutter	Will Cut Gears From:	Number of Cutter	Will Cut Gears From:
1	135 teeth to a rack	5	21 to 25 teeth
2	55 to 134 teeth	6	17 to 20 teeth
3	35 to 54 teeth	7	14 to 16 teeth
4	26 to 34 teeth	8	12 to 13 teeth

(The regular cutters listed sbove are used ordinarily. The cutters listed below (an intermediate series baving half-numbers) may be used when greater accuracy of tooth shape is essential in cases where the number or teeth is between the number for which the regular cutters are intended.)

Table 8.6

Number of Cutter	Will Cut Gears From:	Number of Cutter	Will Cut Gears From:
1—1/2	80 to 134 teeth	5—1/2	19 to 20 teeth
2—1/2	42 to 54 teeth	6—1/2	15 to 16 teeth
3—1/2	30 to 34 teeth	7—1/2	13 teeth
4—1/2	23 to 25 teeth		

8.7 Milling Machine Adjustments

8.7.1 Vertical Milling Machine

(1) Adjustments (Fig. 8.20).

(a) Proper gib adjustment procedures must be done after 40 hours on new mills.

(b) Each 700 and 800 series of mills have three gibs. One at the front dovetail of the table, one on the left dovetail of the saddle, and one on the left dovetail of the knee. Each gib is supplied with two lock or adjustment screws. The table gib has a lock screw on the right front of the saddle and the adjusting screw on the left front of the saddle.

(c) The saddle gib is at the rear of the saddle on the left side, while the adjusting screw is at the front of the saddle on the left side. The knee gib lock screw is on the bottom of the knee on the left side and the adjusting screw is on the top on the left side.

(d) To adjust the table gib, loosen the table gib lock screw several turns and tighten the adjusting screw on the opposite side of the table until the gib is pressing against the table dovetail. Tighten the lock screw. (Do not tighten the lock screw too tight as it distorts the gib.) Run the table back and forth and check the table for drag. To adjust the saddle and knee, use the same procedures as above.

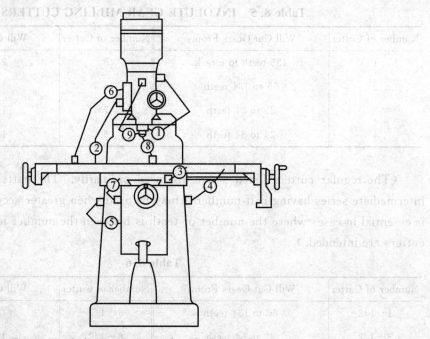

Fig. 8.20   Indicatior Positions for Checking Adjustments

(2) Adjustments With The Dial Indicator.

(a) When checking the gibs with a dial indicator, the following checks should be made.

(b) With the dial indicator mounted, as in Position 3, the table can be tested for looseness by pulling back and forth on the end of the table. Anything over 0.0015 - inch is too much and requires the gibs to be adjusted, also the table should snap back to the "0" reading each time after the table is released.

(c) To check the saddle gib, the indicator should be mounted as in Position 7 and the same tolerance should exist here.

(d) The knee gib will be checked as shown, with the dial indicator in Position 5, by grasping the table and lifting up and pushing down. The reading of deflection here should not be more than 0.0003 of an inch.

(e) As a final check, set the dial indicator on Position 2 and run the table to its extreme right and left positions. The indicator runout should not be more than 0.0015 of an inch.

(3) Quill Feed Clutch.

Adjustment of this clutch is as follows:

(a) In the rear of the head, between the head and the adapter, is a hex nut with a 10-24×1/2 inch socket head cap screw, used for a lock. Unscrew this lock screw until you can

Chapter 8 Milling and Milling Machines

rotate it freely with your fingers.

(b) The minimum clutch tension is preset at the factory. If more tension is desired, rotate the hex nut to 180° from its present position, and relock the NO. 10-24 socket head screw.

(c) It may also be desirable to adjust the travel of the clutch plunger:

①This adjustment is made by means of the 1/4-20 * 3/4 socket set screw immediately behind the feed cam housing.

②With the clutch disengaged, tighten the set screw (clockwise) while moving the quill down by means of the hand lever until a little roughness can be felt.

(4) Troubleshooting.

NOTE

Ordinarily trouble will not manifest itself except when actually working with the machine.

(a) Slide ways forking hard or binding.

• Cause - The gibs are out of adjustment, they are either too tight or too loose. This causes the gib to "wedge."

• Remedy - Adjust the gibs.

• Cause - There is dirt in the slide ways.

• Remedy - Wash out slide ways with light oil.

(b) Chatter or vibration when cutting.

• Cause - Dirt in the spindle taper, causing a bad fit between the tool holder shank and the spindle taper.

• Remedy - Clean the spindle taper and the shank of the tool holder.

• Cause - Faulty shank on the tool holder.

• Remedy - Replace the shank or dress off the burrs, if they are due to nicks.

• Cause - Gibs poorly adjusted on the slide ways, or dirty.

• Remedy - Clean and adjust the gibs as required.

• Cause - Work improperly clamped to the table of the milling machine.

• Remedy - Check for rocking or movement, and correct by proper clamping.

• Cause - Improper ground cutting tool.

• Remedy - Replace or regrind the tool.

• Cause - Hard spot at the splice of the drive or the worm belts.

• Remedy - Replace the belts.

• Cause - Spindle quill worn in the quill head.

• Remedy - Tighten the quill head lock slightly.

• Cause - Incorrect spindle speed, table feed, or both.

• Remedy - Ordinarily, increase the spindle speed and/or increase or decrease the feed to break up the vibration period. Experiment by using the hand feed to feed the table.

• Cause - Drive pulleys are worn in the grooves or loose on the shafts.

- Remedy - Replace the pulleys.

(c) Boring or milling out of square or at an angle.
- Cause - The head is not properly aligned with the table.
- Remedy - Check the head for alignment and correct.
- Cause - Work improperly set up; i.e., not square or flat.
- Remedy - Check and re-align the work.

(d) Failure to hold the center distance when locating for boring.
- Cause - Failure to take back-off tension on the lead screw after coming up to the dial indicator reading, causing the table to "creep"; or failure to lock up the slide ways with the same amount of tension after moving the table to a new position.
- Remedy - Take the back-off tension off from the lead screw after coming up to the indicator reading and lock the table in position.

8.7.2 Plain Milling Machine-knee Type

(1) General.

The spindle bearings used in the spindle head are the taper roller type and have been properly adjusted for average conditions before leaving the factory. They should not require readjustment before the machine is to be used. If desired, the end play in the spindle bearings may be checked after a few months of operation in the following manner: Using a lead or composition hammer, gently tap the face of the spindle until all of the play is taken up towards the rear of the machine. Place a dial indicator against the face of the spindle, and then tap the spindle shaft forward from the rear. If the play exceeds 0.001 inch, adjustment may be made as follows:

CAUTION

Do not attempt to take up the spindle bearings without a thorough knowledge of the bearing adjustments and the operating conditions.

(2) Adjustment.
(a) Remove the rear cap from the cutter head.
(b) Loosen the set screws in the bearing head.
(c) Loosen or tighten the nut as necessary to secure the desired adjustment.
(d) Tighten the set screw and reassemble the rear cap on the cutter head.

(3) Checking Gib Adjustments.
(a) Gib adjustments should-only be made by those who are thoroughly acquainted with the operation of the machine. In general, all of the gibs should be tight enough to eliminate any and all play, but not so tight that there will be a heavy drag on the working parts. Gibs that are too loose will result in inaccurate work; gibs that are too tight will cause severe wear and strain on the operating mechanisms.

(b) The gibs are properly adjusted at the factory. When readjustment becomes necessary, proceed as follows: The table gib is adjusted by means of two shouldered screws located on each end of the saddle. By first loosening one and then tightening the other, the taper gib may be adjusted as needed. The saddle and knee gibs are adjusted in a similar manner. (c) The ram gib is adjusted by two adjusting screws. The front and the end screw are the adjusting screws. To adjust the gib, the ram stops must first be removed. These stops are located on the bottom of the ram. To remove the stops, loosen the set screws, which are located on the front and back end of the column, and turn both screws on the gib an equal amount.

After proper adjustment has been accomplished, retighten the set screws in the column to hold the adjustment and then replace the ram stops.

(4) Table Feed Screw.

Backlash in the table feed screw is adjusted by an adjustable bronze feed screw nut that is located at the left hand end of the saddle. This nut is located in the saddle mechanism just above the saddle binder. To make any desired adjustment, first loosen the check nut and insert a pin in any of the many holes around the flange of the nut. Then turn the screw in either direction until the backlash is from 0.002-inch to 0.005-inch. After you have completed this adjustment, tighten the check nut.

(5) Saddle Feed Screw.

The saddle feed screw is adjusted by means of a bronze adjusting nut. This adjusting nut is located in the rear end of the bracket which is used to carry the screw, located under the saddle. To adjust the feed screw, remove the plate from the left side of the knee. Then loosen the two Allen screws that holds the bronze nut. Insert a pin into one of the holes in the circumference of the nut and turn it until the backlash is from 0.002-inch to 0.005-inch. After the correct adjustment has been made, tighten the two Allen screws and replace the cover plate.

8.8 Milling Cutters

There are different types of milling machine cutters. Some cutters can be used for several operations, others can be used for only one operation. Some cutters have straight teeth, others have helical teeth. Some cutters have mounting shanks, others have mounting holes. The machine operator must decide which cutter to use. To make this decision, he must be familiar with various types of cutters and their uses.

Standard milling cutters are made in many shapes and sizes for milling both regular and irregular shapes. Various cutters designed for specific purposes also are available.

Milling cutters generally take their names from the operation which they perform.

Those commonly recognized are:

(1) plain milling cutters of various widths and diameters, used principally for milling flat surfaces which are parallel to the axis to the cutter;

(2) angular milling cutters, designed for milling V-grooves and the grooves in reamers, taps, and milling cutters;

(3) face milling cutters, used for milling flat surfaces at right angles to the axis of the cutter;

(4) forming cutters, used for the production of surfaces with some form of irregular outline.

Milling cutters may also be classified as arbor-mounted, or shankmounted. Arbor-mounted cutters are mounted on the straight shanks of an arbor. The arbor is then inserted into the milling machine spindle.

Milling cutters may have straight, right-hand, left-hand, or staggered teeth. Straight teeth are parallel to the axis of the cutter. If a helix angle twists in a clockwise direction, the cutter has right-hand teeth. If the helix angle twists in a counterclockwise direction, the cutter has left-hand teeth. The teeth on staggered-tooth cutters are alternately left-hand and right-hand.

1. Milling Cutter Nomenclature

Fig. 8.21 shows two views of a common milling cutter with its parts and angles identified. These parts and angles are common to all types of cutters in some form.

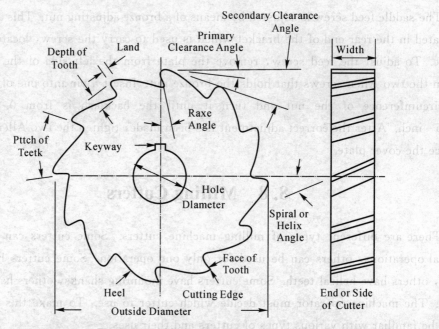

Fig. 8.21 Milling Cutter Nomenclature

(1) Pitch. The pitch refers to the angular distance between like parts on the adjacent teeth. The pitch is determined by the number of teeth.

(2) Face of tooth. The tooth face is the forward facing surface of the tooth which forms the cutting edge.

(3) Cutting edge. The cutting edge is the angle on each tooth which performs the cutting.

(4) Land. The land is the narrow surface behind the cutting edge of each tooth.

(5) Rake angle. The rake angle is the angle formed between the face of the tooth and the centerline of the cutter. The rake angle defines the cutting edge and provides a path for chips that are cut from the workpiece.

(6) Primary clearance angle. The primary clearance angle is the angle of the land of each tooth, measured from a line tangent to the centerline of the cutter at the cutting edge. This angle prevents each tooth from rubbing against the workpiece after it makes its cut.

(7) Secondary clearance angle. The secondary clearance angle defines the land of each tooth and provides additional clearance for the passage of cutting oil and the chips.

(8) Hole diameter. The hole diameter determines the size of arbor that is necessary to mount the milling cutter.

(9) Keyway. A keyway is present on all arbor-mounting cutters for locking the cutter to the arbor.

(10) Spiral or helix angle.

(a) Plain milling cutters that are more than 3/4 inch in width are usually made with spiral or helical teeth.

(b) A plain spiral-tooth milling cutter produces a better and smoother finish, and requires less power to operate.

(c) A plain helix-tooth milling cutter is especially desirable where an uneven surface or one with holes in it is to be milled.

(11) Types of teeth. The teeth of milling cutters are either right-hand or left-hand, viewed from the back of the machine. Right-hand milling cutters cut when rotated clockwise; left-hand milling cutters cut when rotated counterclockwise.

(a) Saw Teeth. Saw teeth similar to those shown in Fig. 8.23 on the previous page are either straight or helical in the smaller sizes of plain milling cutters, metal slitting saw milling cutters, and end milling cutters. The cutting edge is usually given about 5° primary clearance angle. Sometimes the teeth are provided with offset nicks which break up the chips and make coarser feeds possible.

(b) Formed Teeth. Formed teeth are usually specially made for machining irregular surfaces or profiles. The possible varieties of formed-tooth milling cutters are almost unlimited. Convex, concave, and corner-rounding milling cutters are of this type. Formed cutters are sharpened by grinding the faces of the teeth radially. Repeated sharpenings are possible without changing the contour of the cutting edge.

(c) Inserted Teeth. Inserted teeth are blades of high-speed steel inserted and rigidly held in a blank of machine steel or cast iron. Different manufacturers use different methods of holding the blades in place. Inserted teeth are more economical and convenient for large-size cutters because of their reasonable initial cost and because worn or broken blades can be replaced more easily and at less cost.

2. Kinds of Milling Cutters

(1) Plain Milling Cutter (Fig. 8.22).

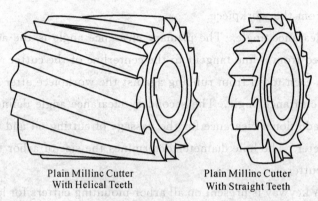

Fig. 8.22 Plain Milling Cutters

The most common type of milling cutter is known as a plain milling cutter. It is merely a metal cylinder having teeth cut on its periphery for producing a flat horizontal surface (or a flat vertical surface in the case of a vertical spindle machine). When the cutter is over 3/4 inch wide, the teethare usually helical, which gives the tool a shearing action which requires less power, reduces chatter, and produces a smoother finish. Cutters with faces less than 3/4 inch wide are sometimes made with staggered or alternate right-and left-hand helical teeth. The shearing action, alternately right and left, eliminates side thrust on the cutter and arbor. When a plain milling cutter is considerably wider than its diameter, it is often called a slabbing cutter; slabbing cutters may have nicked teeth that prevent formation of large chips.

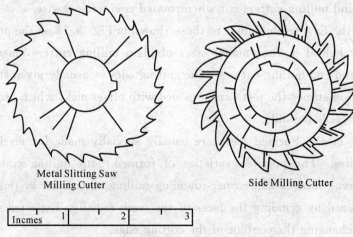

Fig. 8.23 Side and Metal Slitting Saw Milling Cutters

(2) Metal Slitting Saw Milling Cutter (Fig. 8.23).

The metal slitting saw milling cutter is essentially a very thin, plain milling cutter. It is ground slightly thinner toward the center to provide side clearance. It is used for metal sawing and for cutting narrow slots in metal.

(3) Side Milling Cutters.

Essentially plain milling cutters with the addition of teeth on one or both sides.

A side milling cutter has teeth on both sides and on the periphery. When teeth are added to one side only, the cutter is called a half-side milling cutter and is identified as being either a right-hand or left-hand cutter. Side milling cutters are generally used for slotting and straddle milling.

Interlocking tooth side milling cutters and staggered tooth side milling cutters (Fig. 8.24 on the following page) are used for cutting relatively wide slots with accuracy. Interlocking tooth side milling cutters can be repeatedly sharpening without changing the width of the slot that will be machined. After each sharpening, a washer is placed between the two cutters to compensate for the ground-off metal. The staggered tooth cutter is the most efficient type used for milling slots where the depth exceeds the width.

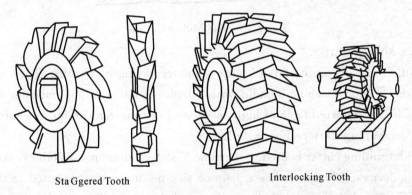

StaGgered Tooth   Interlocking Tooth

Fig. 8.24 Side Milling Cutters

(4) End Milling Cutters.

End milling cutters, also called end mills, have teeth on the end as well as the periphery (Fig. 8.25 on the following page). The smaller end milling cutters have shanks for chuck mounting or direct spindle mounting. Larger end milling cutters (over 2 inches in diameter) are called shell end milling cutters and are mounted on arbors like plain milling cutters. End milling cutters are employed in the production of slots, keyways, recesses, and tangs. They are also used for milling angles, shoulders, and the edges of workpieces.

End milling cutters may have straight or spiral flutes. Spiral flute end milling cutters are classified as left-hand or right-hand cutters, depending on the direction of rotation of the flutes. If they are small cutters, they may have either a straight or tapered shank.

Several common types of end milling cutters are illustrated in Fig. 8.27. The most common end milling cutter is the spiral flute end milling cutter, which contains four flutes. Two fluted end milling cutters are used for milling slots and keyways where no drilled hole is provided for starting the cut. These cutters drill their own starting holes. Straight flute end milling cutters are generally used for milling soft or tough materials, while spiral flute cutters are used mostly for cutting steel.

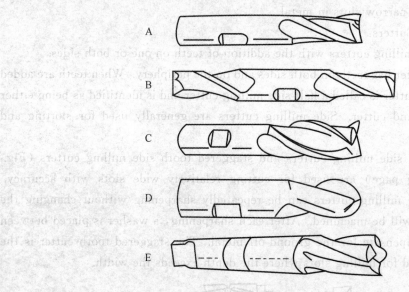

Fig. 8.25 End Milling Cutters

(5) Face Milling Cutter.

Face milling cutters are cutters of large diameter having no shanks. They are fastened directly to the milling machine spindle with adapters. Face milling machine cutters are generally made with inserted teeth of high-speed steel or tungsten carbide in a soft steel hub.

(6) T-Slot Milling Cutter (Fig. 8.26).

The T-slot milling cutter is used to machine T-slot grooves in worktables, fixtures, and other holding devices. The cutter has a plain or side milling cutter mounted to the end of a narrow shank. The throat of the T-slot is first milled with a side or end milling cutter and the headspace is then milled with the T-slot milling cutter.

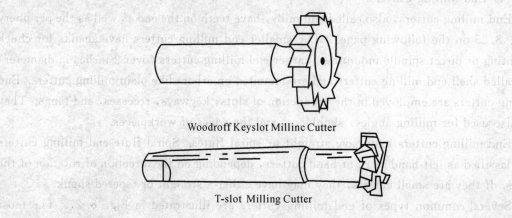

Fig. 8.26 T-slot Milling Cutter

(7) Woodruff Keyslot Milling Cutter.

The woodruff keyslot milling cutter (Fig. 27) is made in straight-shank, tapered-shank, and arbormounted types. The most common cutters of this type, under 1 1/2 inches in diameter, are provided with a shank. They have teeth on the periphery and slightly concave sides to provide clearance. These cutters are used for milling semicylindrical keyways in shafts.

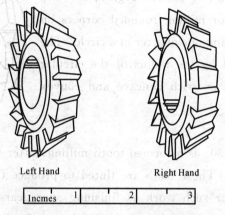

Fig. 8.27 Single-angle Milling Cutters

(8) Angle Milling Cutters.

The angle milling cutter has peripheral teeth which are neither parallel nor perpendicular to the cutter axis. Common operations performed with angle cutters are cutting teeth in ratchet wheels, milling dovetails, and cutting V-grooves. Angle cutters may be single-angle milling cutters (Fig. 8.28) or double-angle milling cutters. The single-angle cutter contains side-cutting teeth on the flat side of the cutter. The angle of the right or left cutter edge is usually 30°, 45°, or 60°. Double-angle cutters have included the angles of 45°, 60°, and 90°.

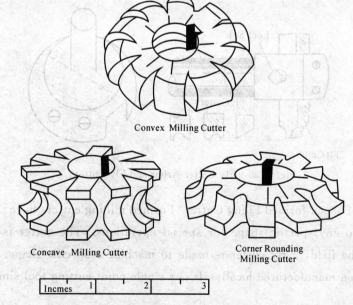

Fig. 8.28 Cconcave, Convex and Corner Rounding Milling Cutters

(9) Concave and Convex Milling Cutters.

Concave and convex milling cutters (Fig. 8.29 on the following page) are formed tooth cutters shaped to produce concave and convex contours of one-half circle or less. The size of the cutter is specified by the diameter of the circular form the cutter produces.

(10) Corner-rounding Milling Cutter.

The corner-rounding milling cutter (Fig. 8.29) is a formed tooth cutter used for milling rounded corners on workpieces up to and including one-quarter of a circle. The size of a cutter is specified by the radius of the circular form the cutter produces, as with concave and convex cutters.

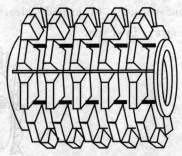

Fig. 8.29 Gear hob

(11) Gear Hob.

The gear hob (Fig. 8.30) is a formed-tooth milling cutter with helical teeth arranged like the thread on a screw. These teeth are fluted to produce the required cutting edges. Hobs are generally used for such work as finishing spur gears, spiral gears, and worm wheels. They may also be employed for cutting ratchets and spline shafts.

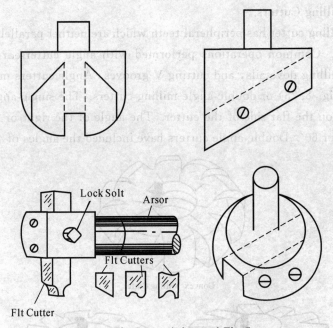

Fig. 8.30 Fly Cutter Arbor and Fly Cutters

(12) Special Shaped-formed Filing Cutter. Formed milling cutters have the advantage of being adaptable to any specific shape for special operations. The cutter is made for each specific job. In the field, a fly cutter is made to machine a specific shape. The fly cutter (figure 32) is often manufactured locally. It is a single-point cutting tool similar in shape to

a lathe or shaper tool. It is held and rotated by a fly cutter arbor. The cutter can be ground to almost any shape, form, or contour that is desired.

The cutter can be sharpened many times without destroying the shape of the cutter or the cut being made. There will be a very limited number of times when a special formed cutter will be needed for cutting or boring operations, this is why a fly cutter is the most practical cutter to use in this type of situation.

3. Selection of Milling Cutters

The following factors should be considered in the choice of milling cutters:

(1) Type of machine to Be used. High-speed steel, stellite, and cemented carbide cutters have the distinct advantage of being capable of rapid production when used on a machine that can reach the proper speed.

(2) Method of Folding the Workpiece. For example, 45° angular cuts may either be made with a 45° single-angle milling cutter while the workpiece is held in a swiveled vise, or with an end milling cutter while the workpiece is set at the required angle in a universal vise.

(3) Hardness of the material to be cut. The harder the material, the greater will be the heat that is generated during the cutting process. Cutters should be selected for their heat-resisting properties.

(4) Amount of material to be removed. A course-toothed milling cutter should be used for roughing cuts, whereas a finer toothed milling cutter may be used for light cuts and finishing operations.

(5) Number of pieces to be cut. For example, when milling stock to length, the choice of using a pair of side milling cutters to straddle the workpiece, a single-side milling cutter, or an end milling cutter will depend upon the number of pieces to be cut.

(6) Class of work being done. Some operations can be accomplished with more than one type of cutter, such as in milling the square end on a shaft or reamer shank. In this case, one or two side milling cutters or an end milling cutter may be used. However, for the majority of operations, cutters are specially designed and named for the operation they are to perform.

(7) Rigidity and size of the workpiece. The milling cutter used should be small enough in diameter so that the pressure of the cut will not cause the workpiece to be sprung or displaced while being milled.

(8) Size of the milling cutter. In selecting a milling cutter for a particular job, it should be remembered that a small diameter cutter will pass over a surface in a shorter time than a large diameter cutter fed at the same speed. This fact is illustrated in Fig. 8.31.

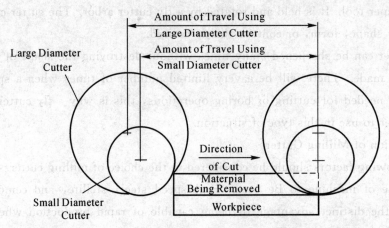

Fig. 8.31 Effect of Milling Gutter Diameter on Workpiece Travel

4. Care and Maintenance of Milling Cutters

The life of a milling cutter can be greatly prolonged by intelligent use and proper storage. General rules for the care and maintenance of milling cutters are given below:

(1) New cutters received from stock are usually wrapped in oilpaper which should not be removed until the cutter is to be used.

(2) Care should be taken to operate the machine at the proper speed for the cutter that is being used; excessive speed will cause the cutter to wear rapidly from overheating.

(3) Whenever practicable, the proper cutting oil should be used on the cutter and the workpiece during the operation, since lubrication helps prevent overheating and consequent cutter wear.

(4) Cutters should be kept sharp, because dull cutters require more power to drive them and this power, being transformed into heat, softens the cutting edges. Dull cutters should be marked as such and set aside for grinding.

(5) A cutter should never be operated backward because, due to the clearance angle, the cutter will rub, producing a great deal of frictional heat. Operating the cutter backward may result in cutter breakage.

(6) Care should be taken to prevent the putter from striking the hard jaws of the vise, chuck, clamping bolts, or nuts.

(7) A milling cutter should be thoroughly cleaned and lightly coated with oil before storing.

(8) Cutters should be placed in drawers or bins in such a manner that their cutting edges will not strike each other. Small cutters that have a hole in the center should be hung on hooks or pegs, large cutters should be set on end. Tapered and straight shank cutters may be placed in separate drawers, bins, or racks provided with suitable sized holes to receive the shanks.

Chapter 8 Milling and Milling Machines

Exercises

1. How are knee-type milling machines characterized?
2. What is the basic difference between a universal horizontal and a plain horizontal milling machine?
3. Which component holds and drives the various cutting tools on the knee-type milling machine?
4. Which component on the ram-type milling machine aligns the outer end of the arbor with the spindle?
5. How are milling machines classified or identified?
6. What is the most common means of fastening the arbor in the milling machine spindle?
7. What is another name for the indexing head?
8. When clamping workpieces to the table, what is used to protect the surfaces of the workpiece?
9. Workpieces mounted between centers are prevented from backlash by what type device?
10. What is the ratio on a standard indexing head?
11. Direct indexing plates usually have how many holes?
12. Face milling operations require the flat surface to be machined at what angles to the axis to the cutter?
13. What determines the cutting speed for a milling operation?
14. What is angular milling?
15. Which teeth do most of the cutting during a face milling operation?
16. What is the required time allocated for proper gib adjustment on a new milling machine?
17. When making knee gib adjustments with a dial indicator, the deflection should not be more than how many thousandths of an inch?
18. When the gibs are adjusted too tightly, what is the end result?
19. Which type milling cutter requires less power to operate and produces a smoother finish?
20. How do milling cutters receive their names?
21. Pitch on a milling cutter refers to the _____.
22. What is the most common type of milling cutter?
23. Which type cutters are used for cutting relatively wide slots with accuracy?

Chapter 9 Benchwork Tools, Drilling, Cutting, Sharpening

Teaching Objectives

In this chapter, students should get familiar with bench tools such as files, handsaws, hammers, and also basic drilling, cutting, and sharpening procedures.

9.1 Brief Introduction

Even in the time of CNC technology it is important to know how to do bench work using different hand tools, because still today bench work plays a big rule in machine maintenance or in metal fabrication. Benchwork skills give a workman the flexibility which is impossible to attain with a machine. In ancient China, virtually every product is a labor of benchwork skills.

9.2 Benchwork Tools

Roughly, benchwork tools incudes laying out tools, files, saws, chisels, workbench, vise, hammers and other tools, however, this chapter will only focus on several of these tools which may come into use in our workshop practice.

9.3 Work Bench

- The workbench should be sturdy and when possible fixed with the shop floor (Fig. 9.1).

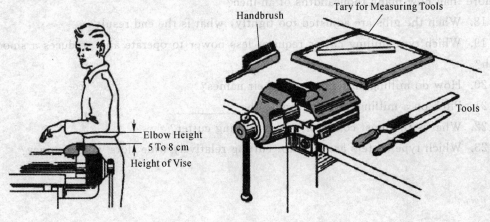

Fig. 9.1

- It is advisable to use wood for the bench board.
- The height of the workbench should depend on the height of the craftsman.
- Keep the workbench clean. Put only the tools necessary for the work on it.
- Measuring tools should be separated from the other tools. Place them accurately on the wooden tray board.

9.4 Bench Vise

- The base of a bench vise is normally made of cast iron. The jaws are hardened. Clamping soft workpieces requires covering the jaws with an aluminum sheet cover.
- The size of the bench vise is measured by the width of the jaws and the maximum opening between the jaws (Fig. 9.2).

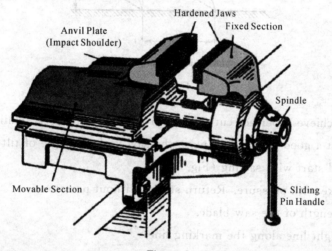

Fig. 9.2

- There are different types of bench vises available: With or without an anvil plate, with a pipe clamping device, machine vise for drill press, and adjustable in any position within 360 degree.

9.5 Hand Hacksaw

A hand hacksaw mainly serves to separate materials and also to produce grooves and slits. By moving the saw in the direction of cut (cutting motion) with simultaneous pressure on the saw (cutting pressure), the teeth penetrate into the material and remove chips (Fig. 9.3).

There are different hacksaw blades, depending on the metal to be cut (Fig. 9.4), available:

Coarse: for soft materials appr. 14 teeth per inch.
Medium: for normal material appr. 22 teeth per inch.
Fine: for hard material appr. 32 teeth per inch.

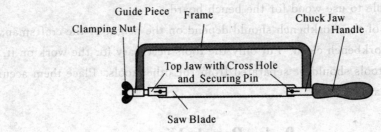

Fig. 9.3

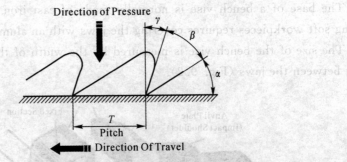

Fig. 9.4

In order to achieve a perfect cut, file with a triangular file a small notch beside the marking line to get a good start, then place the saw with an angle of tilt (as shown in the picture below) and start with sawing (Fig. 9.5).

Forward stroke with pressure. Return stroke without pressure.

Use the full length of the saw blade.

Saw in a straight line along the marking line.

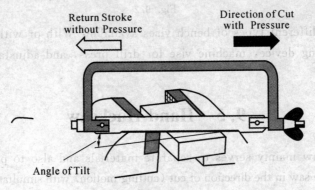

Fig. 9.5

WORK SAFETY:

When sawing through, reduce pressure on hand hacksaw just before the workpiece separate.

9.6 Chisel Tools

In chiseling, the cutting edge of a chisel is driven into a workpiece by impact. A chisel must be harder than the piece being worked. Most chisels are made of alloyed tool steels (Fig. 9.6).

- Wedge angle for soft materials: 30 – 50 degree;
- For mild steel: 60 – 70 degree;
- For alloyed steels: 70 – 80 degree

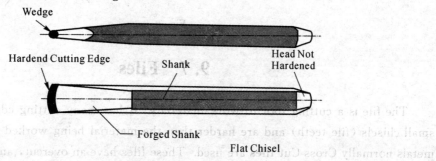

Fig. 9.6　Flat Chisel

9.6.1　Work Process

- The workpiece must be properly clamped when chiseling.
- The chisel must be struck on the center of the head, in the direction of the axis of the chisel (Fig. 9.7).
- The correct wedge angle must be maintained when grinding the chisel (measure with an angle gauge). The tool must be cooled frequently when sharpening, so that it does not lose its temper (Fig. 9.8).

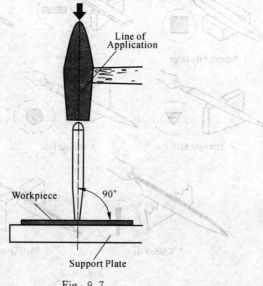

Fig. 9.7

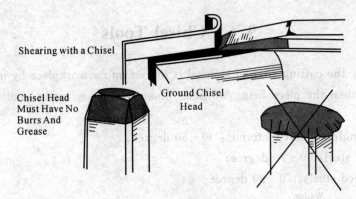

Fig. 9.8

9.7 Files

The file is a cutting tool to work materials. It has many cutting edges which are like small chisels (file teeth) and are harder than the material being worked upon. For cutting metals normally Cross-Cut files are used. These files have an overcut, and an upcut. When using a file, several cutting wedges always act at the same time.

- To file different materials there are various files available, such as smooth-cut, second-cut, and bastard cut.
- The length of the file body normally used is between 100 mm and 350 mm.
- The file handle is either from wood or from plastic.

9.7.1 Types of Files (Fig. 9.9)

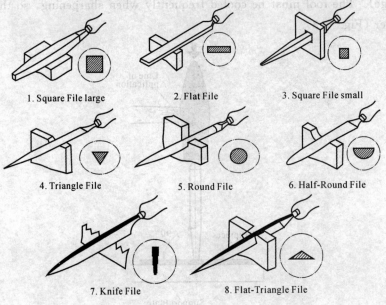

Fig. 9.9

9.7.2 Selection of Files

• Based on metals: for soft metals, such as copper, aluminum, select rough files, for hard metals, select fine files.

• Based on precision requirement of workpiece: Select rough files for low-requirement files, and fine files for high-requirement files.

• Based on the shape of workpiece (Fig. 9.10).

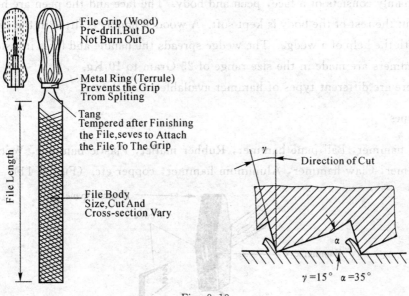

Fig. 9.10

9.7.3 File Handling

• Clamp the workpiece asclose as possible to the jaws of the vise. Use protective jaws (Aluminum) to protect the workpiece.

• Start with arough file for removing more material then take a smooth file to reach a good surface.

• Forward strokewith pressure; Return stroke without pressure.

• Move with the filecrosswise to control the area of filing.

• Clean the file from time to time (especially smooth files) with a wire brush to prevent messy finishes.

• Never work with a filewithout a file grip.

• Make sure that the file grip is properly attached, that it has the right dimension and that it is not splitted.

• The length of the file body normally used is between 100 mm and 350 mm.

• The file handle is either from wood or from plastic.

In the beginning, apply more pressure on the left hand, and more pushing force on the right hand, then gradually change the pressuring force onto the right hand and the pushing force onto the left hand.

9.8 Hammer

- A hammer is used nearly in every operation related to metal works.
- They are made of cast steel or carbon steel.
- It mainly consists of a face, peen and body. The face and the peen are hardened and tempered but the rest of the body is kept soft. A wooden handle is fitted in the eyehole of the hammer with the help of a wedge. The wedge spreads the handle and fixes it inside the hole.
- Hammers are made in the size range of 25 Gram to 10 Kg.
- There are different types of hammer available:

Hammer Types

Fitters hammer, Ball pane hammer, Rubber mallet, Plastic hammer, Wood hammer, Sledge hammer, Claw hammer, Aluminum hammer, copper etc. (Fig. 9.11).

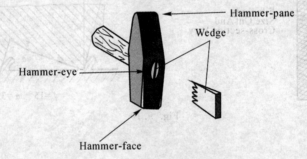

Fig. 9.11

Safety:

- The handle of a hammer should be dry and not greasy
- The surface of a handle should be smooth
- The face of a hammer should not be spotted, if it so then make it smooth by grinding
- Hold the hammer handle always nearer to its tail end.

9.9 Metal Cutting

A large portion of manufacturing operations in the world consists of machining metal to size and shape. To be competitive, it is important that machining operations be as cost-efficient as possible. This requires a good knowledge of metals, cutting tools, and machining conditions and processes (Fig. 9.12).

Chapter 9 Benchwork Tools, Drilling, Cutting, Sharpening

Table 9.1 Classification of Metal Cutting Processes

Hand Cutting Processes	Machine Cutting Processes
• Filing	• Drilling
• Chiseling	• Hacksawing
• Hand Hacksawing	• Turning
• Shearing	• Milling
• Hand Tapping	• Grinding
• Die-Tapping	• Shaping
• Hand Reaming	• Machine Threading
	• Machine Reaming

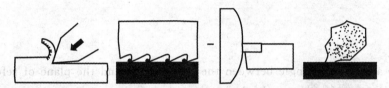

Fig. 9.12 Chiseling, Sawing, Turning and Grinding

9.9.1 Angles of tools

• What is common to all cutting tools is the wedge shape.

• To cut metals, the tool must be wedge-shaped, be resistant to abrasion and tenacious.

• For different cutting operations there is a need for different tool angles.

• Cutting tools with small wedge angles penetrate the material more easily but also tend to break off more easily if the material is hard (Fig. 9.13).

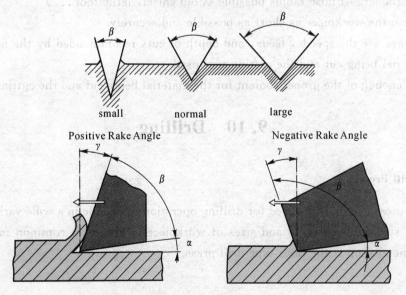

Fig. 9.13

Wedge Angle:

The wedge angle must suit to the material being worked.
- The smaller the wedge angle is, the lower the expenditure of force.
- The harder the material, the larger the wedge angle should be chosen.

Clearance Angle:

The clearance angle is the angle between the flank of the tool and the surface being cut. Friction and heating depend upon this angle. The angle should be chosen as such that the tool could cut freely.
- Soft materials require a larger clearance angle because they generate more heat and friction.

Rake Angle:

The rake angle is the angle between the cutting face and the plane of reference of the tool, an imaginary surface perpendicular to the cut surface.

The rake angle influences the chip formation.
- Large angle: good chip flow, low cutting force
- Small to negative angle: great cutting force, highly robust cutters

9.9.2 Cutting Tool Guideline

- Cutting tools are expensive, therefore, take good care of them.
- Always use sharp cutting tools to ensure an efficient cutting action and accurate work.
- Use the largest nose radius possible (Cold chisel, lathe tool ...)
- Clamp the workpiece as short as possible and securely.
- Always use the speeds, feeds, and depth of cuts recommended by the manufacturer for the material being cut and the cutting tool used.
- Use enough of the proper coolant for the material being cut and the cutting tool used.

9.10 Drilling

9.10.1 Drill Press

A drill press is a machine used for drilling operations available in a wide variety of types and sizes to suit different types and sizes of workpieces. The most common machine type found in a metal shop is the floor-type drill press.

Drill Press Parts:

Although drill presses are manufactured in a wide variety of sizes, all drilling machines contain certain basic parts (Fig. 9.14).

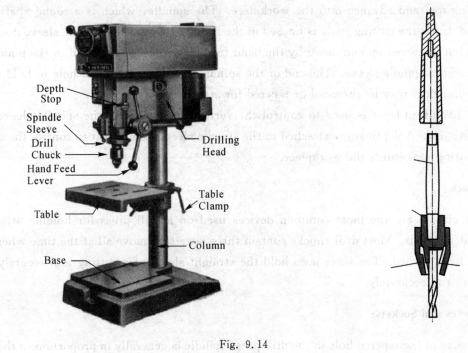

Fig. 9.14

Base:

The base, usually made of cast iron, provides stability for the machine and rigid mounting for the column. The base is usually provided with holes so that it may be bolted to a table or bench to keep it rigid. The slots or ribs in the base allow the work-holding device for the workpiece to be clamped to the base.

Column:

The column is an accurate, vertical, cylindrical post that fits into the base. The table, which is fitted on the column, may be adjusted to any point between the base and head. The head of the drill press is mounted near the top of the column.

Table:

The table, either round or rectangular in shape, is used to support the workpiece to be machined. The table, whose surface is at 90 degree to the column, may be raised, lowered, and swiveled around the column. On some models it is possible to tilt and lock the table in either direction for drilling holes on an angle. Slots are provided in most tables to allow jigs, fixtures, or large workpieces to be clamped directly to the table.

Drilling Head:

The head, mounted close to the top of the column, contains the mechanism to revolve the cutting tool and advance into the workpiece. The spindle, which is a round shaft that holds and drives the cutting tool, is housed in the spindle sleeve. The spindle sleeve does not resolve, but is moved up and down by the hand feed lever that is connected to the pinion on the rack of the spindle sleeve. The end of the spindle may have a tapered hole to hold taper shank tools, or it may be threaded or tapered for attaching a drill chuck.

The hand feed lever is used to control the vertical movement of the spindle sleeve and the cutting tool. A depth stop, attached to the spindle sleeve, can be set to control the depth that a cutting tool enters the workpiece.

Drill Chuck:

Drill chucks are the most common devices used on a drill press for holding straight-shank cutting tools. Most drill chucks contain three jaws that move all at the time when the outer collar is turned. The three jaws hold the straight shank of a cutting tool securely and cause it to run accurately.

Drill Sleeves and Sockets:

The size of the tapered hole in the drill press spindle is generally in proportion to the size of the machine: The larger the machine, the larger the spindle hole. A drill sleeve is used to adapt the cutting tool shank to the machine spindle if the taper on the cutting tool is smaller than the tapered hole in the spindle.

Before a taper shank tool is mounted in a drill press spindle, be sure that the external taper of the tool shank and the internal taper of the spindle are thoroughly cleaned. Align the tang of the tool with the slot in the spindle hole and, with a sharp upward snap, force the tool into the spindle.

Remove a Taper Shank Tool:

A drift, a wedge-shaped tool, is used to remove a taper-shank tool from the drill press spindle. Place a piece of wood under the tool. Insert the drift and sharply strike the end of it with hammer to remove the tool from the drill press spindle.

Twist Drill:

A twist drill is a cutting-tool used to produce a hole in a piece of metal or other material. The most common drill manufactured has two cutting edges (lips) and two straight or helical flutes. The flutes provide the cutting edges with cutting fluid and allow the chips to escape during the drilling operation.

Drill Bit Materials:

High-speed steels drills are the most commonly used drills, since they can be operated at good speeds and the cutting edges can withstand heat and wear.

Cemented-carbide drills, which can be operated much faster than high-speed steel drills, are used to drill hard materials. They can be operated at high speeds and they can withstand higher heat.

Twist Drill Parts and Cutting Angles:

A twist drill may be divided into three main sections (Fig. 9.15):

- Shank: The shank is the part of the drill that fits into a holding device. It may be either straight or tapered.
- Body: The body contains the flutes, margin, and body clearance of the drill.
- Point: Shape and condition of the point are very important to the cutting action of the drill.

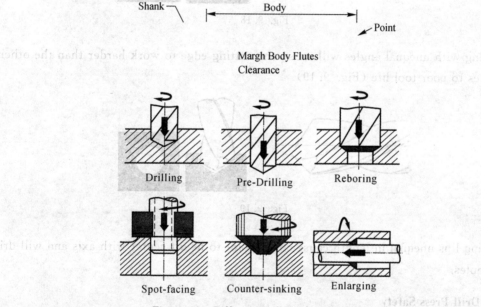

Fig. 9.15 Different Drill Press Operations

9.10.2 Facts and Problems

The most common drill problems encountered are illustrated below (Fig. 9.16).

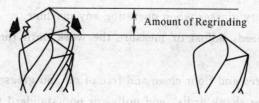

Fig. 9.16

Excessive speed will cause wear at outer corners of drill. This leads to more regrinding of material (Fig. 9.17).

Fig. 9.17

Excessive feed sets up abnormal end thrust that causes breakdown of chisel point and cutting lips. Failure included by this cause will be broken or split drill (Fig. 9.18).

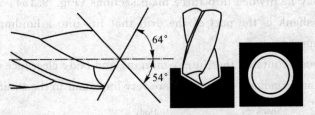

Fig. 9.18

Cutting with unequal angles will cause one cutting edge to work harder than the other. This causes to poor tool life (Fig. 9.19).

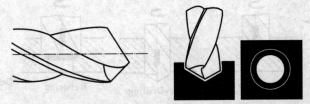

Fig. 9.19

Cutting lips unequal in length cause chisel point to be off center with axis and will drill oversize holes.

9.10.3 Drill Press Safety

- Never wear loose clothing around machinery
- A hair net or a cap protect long hair to prevent it from becoming caught in the revolving parts of the drill press.
- Never wear rings, watches, bracelets or necklaces while working in a machine shop.
- Always wear safety glasses when operating any machine.
- Never set the speed, adjust or measure the work until the machine is completely stopped.
- Keep the work area and floor clean and free of oil and grease.
- Never clamp taper shank drills, end mills, or non-standard tools in a drill chuck.

- Never leave a chuck key in a drill chuck at any time.
- Always use the brush to remove chips.
- Always clamp workpieces when drilling holes larger than - in. (12.7 mm) in diameter.
- When drilling sheet metal, it is necessary to clamp the sheet on a piece of wood.
- Reduce drilling pressure as the drill breaks through the workpiece.
- Always remove the burrs from a hole that has been drilled.

9.11 Cutting Threads with Tap & Dies

Whenever possible, threads should be cut with machines where they can be accurately controlled and the thread cut will be of high quality. Sometimes it may be necessary, due to the size and shape of the workpiece, or because only a few parts are required, to cut the thread with hand tools. Done with care, fairly accurate internal threads can be cut with a tap; external threads can be cut with a die.

Main Parts of a Screw Thread

See Fig. 9.20.

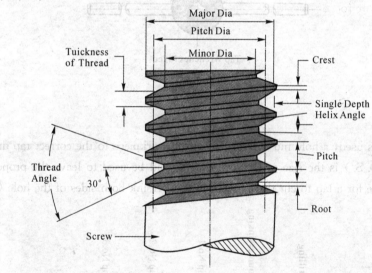

Fig. 9.20 Hand Tapping

Hand Tap

A tap is a cutting tool used to cut internal threads. Normally it's made of high-speed steel (HSS).

Hand taps are usually made in sets of three, because it is better to distribute all the cutting work during the thread-process to three taps.

No. 1 (taper) tap: 1 ring on shank

No. 2 (plug) tap: 2 rings on shank

No. 3 (bottoming) tap: without ring

The most common taps have two or three flutes in order to form the cutting edges, transport the chips out of the hole and give way for the lubricant. The end of the tap is square so that a tap wrench can be used to turn it into a hole (Fig. 9.21).

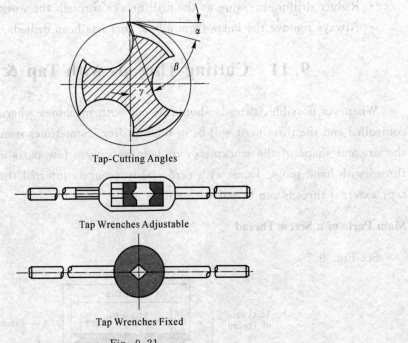

Tap-Cutting Angles

Tap Wrenches Adjustable

Tap Wrenches Fixed

Fig. 9.21

Tapping a Hole

Before a tap is used, a hole must be drilled in the workpiece to the correct tap drill size. The tap drill size (T. D. S.) is the size of the drill that should be used to leave the proper amount of material in the hole for a tap to cut threads. Then countersink both sides of the hole (Fig. 9.22).

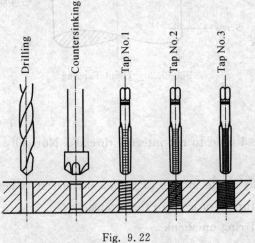

Fig. 9.22

Chapter 9 Benchwork Tools, Drilling, Cutting, Sharpening

If there is no tap drill size chart available, the tap drill size can be easily found by applying simple formulas:

Table 9.2

Inch Threads	Metric Threads
T.D.S. = D in inch - 1/N	T.D.S. = D in mm — P
T.D.S. = tap drill size	T.D.S = tap drill size
D = major diameter of tap	D = major diameter of tap
N = number of threads per inch	P = pitch

Table 9.3 Table of Drill sizes

Metric	Pitch mm	Drill Ø mm	UNC	TPI	Drill Ø mm	UNF	TPI	Drill Ø mm
M 3	0.50	2.5	1/4"	20	5.1	1/4"	28	5.5
M 4	0.70	3.3	5/16"	18	6.6	5/16"	24	6.9
M 5	0.80	4.2	3/8"	16	8.0	3/8"	24	8.5
M 6	1.00	5.0	7/16"	14	9.4	7/16"	20	9.9
M 8	1.25	6.8	1/2"	13	10.8	1/2"	20	11.5
M 10	1.50	8.5	9/16"	12	12.2	9/16"	18	12.9
M 12	1.75	10.2	5/8"	11	13.5	5/8"	18	14.5
M 16	2.00	14.0	3/4"	10	16.5	3/4"	16	17.5
M 20	2.50	17.5	7/8"	9	19.5	7/8"	14	20.4
M 24	3.00	21.0	1"	8	22.25	1"	12	23.25

Working Steps for Hand Tapping (Fig. 9.23)

(1) Select the correct size and type of tap for the job (blind hole or through hole).

(2) Select the correct tap wrench for the size being used.

(3) Use a suitable cutting fluid (No cutting fluid for brass or cast iron).

(4) Place the tap in the hole as near to vertical as possible.

(5) Apply equal down-pressure on both handles, and turn the tap clockwise (for right-hand thread) for about two turns.

(6) Remove the tap wrench and check the tap for squareness. Check at two positions 90 degree to each other.

(7) If the tap has not entered squarely, remove it from the hole and restart it by applying slight pressure in the direction from which the tap leans. Be careful not to exert too much pressure in the straightening process.

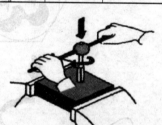
Turn clockwise with light pressure

Check the 90-degree Angle

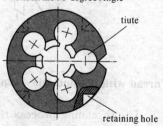
tiute
retaining hole
Adjustable Die

Fig. 9.23

otherwise the tap may be broken.

(8) Turn the tap clockwise one-half turn and then turn it backward about one-quarter of a turn to break the chip. This must be done with a steady motion to avoid breaking the tap.

Threading Dies

A threading die is used to cut external threads on round workpieces. The most common threading dies are the adjustable and solid types. The round adjustable die is split on one side and can be adjusted to cut slightly over or under-sized threads. It is mounted in a die stock, which has two handles for turning the dies onto the work (Fig. 9.24).

The solid die, cannot be adjusted and generally used for re-cutting damaged or oversized threads. Solid dies are turned onto the thread with a special die-stock, or adjustable wrench.

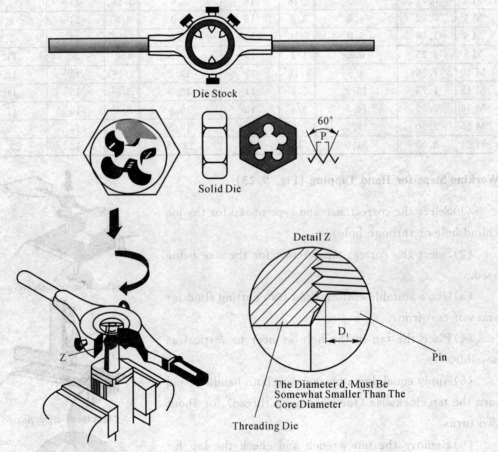

Fig. 9.24

Thread with a Hand Die - Working Steps

The threading process requires the machinist to work carefully to produce usable parts and avoid damage.

The following describes the procedure to be used

(1) Step: Chamfer the end of the workpiece with a file or on the grinder. Consider that a 3/4" thread requires a bolt with an outside diameter of 3/4".

(2) Step: Fasten the workpiece securely in a vise. Hold small diameter work short to prevent it from bending.

(3) Step: Select the proper die and die stock.

(4) Step: Lubricate the tapered end of the die with a suitable cutting lubricant.

(5) Step: Place the tapered end of the die squarely on the workpiece.

(6) Step: Apply down-pressure on both die-stock handles and turn clockwise several turns.

(7) Step: Check the die to see if it has started squarely with the work.

(8) Step: If it is not square, remove the die from the workpiece and restart it squarely, applying slight pressure while the die is being turned.

(9) Step: Turn the die forward one turn, and then reverse it approximately one-half of a turn to break the chip.

(10) Step: Apply cutting fluid frequently during the threading process.

9.12 Sharpening Tools

For some tools it is very important to keep them sharp at all times. Common tools, such as scribers, center punches, chisels, drill bits, tool bits for lathe machine needs to be sharpened every time you feel that they do not cut well.

Bench Grinder or Pedestal Grinder

The bench grinder is used for the sharpening of cutting tools and the rough grinding of metal. Because the work is usually held in the hand, this type of grinding is sometimes called "offhand grinding" (Fig. 9.25).

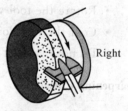
Right

The bench grinder is mounted on a bench while the pedestal grinder being a larger machine, is fastened to the floor. Both types consist of an electric motor with a coarse abrasive grinding wheel for the fast removal of metal, while the other is a fine abrasive wheel for finish grinding.

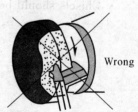
Wrong

Fig. 9.25

- The grinding wheels are normally made of Aluminum-Oxide or Silicon-Carbide. Aluminum-Carbide is used to grind High-Tensile-Strength Materials. Silicon-Carbide is used to grind Low-Tensile-Strength Materials.
- The wheel guards give the necessary protection while grinding.
- The tool rest provide a rest for either the work or hands while grinding.

- The eye shield is an additional protection for the eyes and should be used (Fig. 9.26).

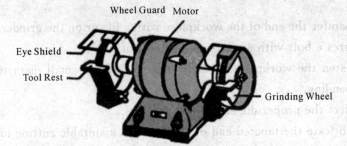

Fig. 9.26

Redressing the grinding wheels

When a grinding wheel is used, several things can happen to it:
- Grooves become worn in the face of the wheel.
- The abrasive grains will lose its cutting action.
- Small metal particles imbed themselves in the wheel, causing it to become loaded or clogged. Use from time to time a disc type dresser or a dressing stone to remove the grooves and the metal particles. This will also re-sharpen the abrasive grains.

Sharpening Tools

Sharpening Scriber and Center Punch
- Scriber and center punch should be ground in the position as shown beside.
- Use the tool rest to rest your hands while bringing the tool in the right position.
- Rotate the tool while grinding.
- Cool the tool down from time to time.
- Do not overheat the metal.

Sharpening Chisel
- Chisels should be ground in the position as shown below (Fig. 9.27).

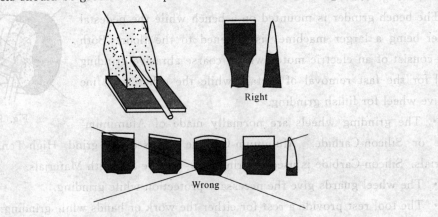

Fig. 9.27

Chapter 9 Benchwork Tools, Drilling, Cutting, Sharpening

- Use the tool rest to rest your hands while bringing the tool in the right position.
- Use the whole grinding wheel while grinding. Move with the tool regularly from the left to the right side and back.
- Cool the tool down from time to time.
- Do not overheat the metal.
- Grind the chisel-point parallel and straight. See also the pictures below.

Safety Precautions:

(1) When switching on the machine, stand beside, because a damaged wheel might burst during acceleration.

(2) Always use safety goggles when grinding.

(3) The tool rest should never have more than 2-3 mm distance to the grinding wheel.

(4) Small workpieces should be held with clamps or other suitable devices.

(5) Keep the metal cool by dipping it frequently in water.

(6) Stand comfortable and don't give too much force to the workpiece because in the case of slip off with the workpiece you will grind your fingers or hand.

(7) While grinding, use only the face of the wheel.

Exercises

1. When sawing a workpiece, how to operate the handsaw?
2. How to thread with a hand die?
3. Rewrite the safety precautions for sharpening.

Chapter 10 A Brief Introduction to Modern Machining

Teaching Objectives

After this chapter, students should know the brief history of modern machining, rough components of CNC machine, CNC programming conventions, and some basics of G code and M code.

10.1 Brief Introduction

Numerical control (NC) is the operation of a machine tool by a series of coded instructions consisting of numbers, letters of the alphabet, and symbols that the machine control unit (MCU) can understand. These instructions are changed into electrical pulses of current that the machine's motors and controls follow to carry out manufacturing operations on a workpiece. The numbers, letters, and symbols are coded instructions that refer to specific distances, positions, functions, or motions, that the machine tool can understand as it machines the workpiece.

10.2 A Short History of Modern Machining

The first commercial NC machines were built in the 1950's, and ran from punched tape. While the concept immediately proved it could save costs, it was so different that it was very slow to catch on with manufacturers. In order to promote more rapid adoption, the US Army bought 120 NC machines and loaned them to various manufacturers so they could become more familiar with the idea. By the end of the 50's, NC was starting to catch on, though there were still a number of issues. For example, g-code, the nearly universal language of CNC we have today, did not exist. Each manufacturer was pushing its own language for defining part programs (the programs the machine tools would execute to create a part) (Fig. 10.1).

Fig. 10.1 1959 CNC Machine: Milwaukee-Matic-II was First Machine with a Tool Changer

10.3 Key Developments

Standard G-Code Language for Part Programs: The origin of G-code dates back to MIT, around 1958, where it was a language used in the MIT Servomechanisms Laboratory. The Electronic Industry Alliance standardized G-code in the early 1960's.

CAD came into its own and started rapidly replacing paper drawings and draftsmen during the 60's. By 1970, CAD was a decent sized industry with players like Intergraph and Computer vision, both of whom I consulted for back in my college days.

Minicomputers like the DEC PDP-8's and Data General Nova's became available in the 60's and made CNC machines both cheaper and more powerful.

By 1970, the economies of most Western countries had slowed and employment costs were rising. With the 60's, having provided the firm technology foundation that was needed, CNC took off and began steadily displacing older technologies such as hydraulic tracers and manual machining.

US companies had largely launched the CNC revolution, but they had been overly focused on the high end. The Germans were the first to see the opportunity to reduce prices of CNC, and by 1979 the Germans were selling more CNC than the US companies. The Japanese repeated the same formula to an even more successful degree and had taken the leadership away from the Germans just one year later, by 1980. In 1971, the 10 largest CNC companies were all US companies, but by 1987, only Cincinnati Milacron was left and they were in 8th place.

More recently, microprocessors have made CNC controls even cheaper, culminating with the availability of CNC for the hobby and personal CNC market. The Enhanced Machine Controller project, or EMC2, was a project to implement an Open Source CNC controller that was started by NIST, the National Institute of Standards and Technology as a demonstration. Sometime in 2000, the project was taken into the public domain and Open Source, and EMC2 appeared a short time later in 2003.

Mach3 was developed by Artsoft founder Art Fenerty as an offshoot of early EMC versions to run on Windows instead of Linux, making it even more accessible to the personal CNC market. Art's company, ArtSoft, was founded in 2001.

Both the EMC2 and Mach3 CNC software programs are alive and thriving today, as are many other CNC technologies.

10.4 Advantages of NC

The major advantages of NC over conventional methods of machine control are as follows:

 • Higher Precision: NC machine tools are capable of machining at very close tolerances, in some operations as small as 0.005 mm.

 • Better Quality: NC systems are capable of maintaining constant working conditions for all parts in a batch thus ensuring less spread of quality characteristics.

 • Higher Productivity: NC machine tools reduce drastically the non-machining time. Adjusting the machine tool for a different product is as easy as changing the computer program and tool turret with the new set of cutting tools required for the particular part.

 • Multi-operational Machining: Some NC machine tools, for example machine centers, are capable of accomplishing a very high number of machining operations thus reducing significantly the number of machine tools in the workshops.

 • Low Operator Qualification: The role of the operation of a NC machine is simply to upload the work piece and to download the finished part. In some cases, industrial robots are employed for material handling, thus eliminating the human operator.

 • Less Time: An easy adjustment of the machine, adjustment requires less time.

10.5 CNC Machining

CNC Machining is a process used in the manufacturing sector that involves the use of computers to control machine tools. Tools that can be controlled in this manner include lathes, mills, routers and grinders. The CNC in CNC Machining stands for Computer Numerical Control.

On the surface, it may look like a normal PC controls the machines, but the computer's unique software and control console are what really sets the system apart for use in CNC machining.

Under CNC Machining, machine tools function through numerical control. A computer program is customized for an object and the machines are programmed with CNC machining language (called G-code) that essentially controls all features like feed rate, coordination, location and speeds. With CNC machining, the computer can control exact positioning and velocity. CNC machining is used in manufacturing both metal and plastic parts.

First a CAD drawing is created (either 2D or 3D), and then a code is created that the CNC machine will understand. The program is loaded and finally an operator runs a test of the program to ensure there are no problems. This trial run is referred to as "cutting air" and it is an important step because any mistake with speed and tool position could result in a scraped part or a damaged machine.

10.6 Advantages of CNC

The process is more precise than manual machining, and can be repeated in exactly the same manner over and over again. Because of the precision possible with CNC Machining, this process can produce complex shapes that would be almost impossible to achieve with manual machining. CNC Machining is used in the production of many complex three-dimensional shapes. It is because of these qualities that CNC Machining is used in jobs that need a high level of precision or very repetitive tasks.

The composition of CNC machine tools

1. Input and Output Device

Before processing the workpiece, there is a need to devise a feasible working procedure, and send the procedure into numerical control device, when processing, the numerical control device then change the procedure into machine-readable directions and then send via output device to the lathe (Fig. 10.2).

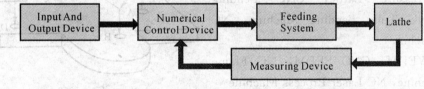

Fig. 10.2

2. Numerical Control Device

The numerical control device is the core of a NC machine, whose function is to code the input parameters of a workpiece, then send machine-readable directions to lathe. Such directions may include:

• Spindle control (start, end, rotation speed, rotation direction, etc.)

• Feeding control (position, trace, speed)

• Offset control (size of tools, transmission interval, error offset, etc.)

• Auxiliary functions (cooling, lubricating, defragment, tool changing, automatic diagnosis, display, intercom)

3. Feeding system

The feeding system will receive directions from NC centre and execute movements like move the tools or working table to or from the workpiece, and adjust the precision of them. The system include Feeding motors, amplifiers, servo control. Its function will have a lot to do with the precision and efficiency of NC lathe.

4. Lathe

The lathe usually include the base, the main spindle, feeding system, etc. NC lathe usually adopt ball screw, rolling guideway to promote transmission efficiency. The machine structure's dynamic stiffness and dampening accuracy are much higher than ordinary lathe, and are highly wear-proof and heatproof.

5. Measuring Device

The measuring part usually adopt speed-measuring motor and photoelectric coded disk to measure the feeding motor's movement, and indirectly measure out the actual shift. The shift will be used to compare with the directions, then offset the discrepancy and control the process accuracy.

10.7 Classification of NC Machines

Classified by processing technique

NC Lathe, NC Miller, NC Grinder, NC Reaming & Milling Center, NC Turning Centre, NC Drilling & Grinding Centre.

NC Pressure Lathe, CNC Bending Machine, NC Tube Bending Machine, NC Spinning Machine

CNC WEDM, CNC SEDM, NC Flame Cutting Machine, NC Laser Process Machine

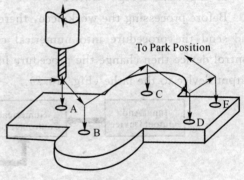

Fig. 10.3

Classified by control system

1. Point to Point Control (Fig. 10.3)

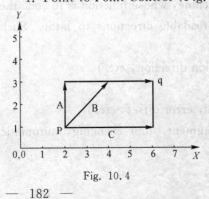

Fig. 10.4

Point to Point Control is the simplest control method, which only control the point where the processing take place, usually used in processing holes.

E. g. NC Drilling machine, NC Coordinate Reaming Machine, NC Punching lathe, etc.

To move the machine table or spindle to a specified position so that machining operations may be performed at that point. The path taken to reach the specific point is not defined, and the movement from one point to the next is non-machining, it is made as rapidly as possible.

2. Straight Line Control (Fig. 10.4)

It is different from the first one. It not only get to the right position, but also machine along the axes when it moves.

3. Contour Control (Fig. 10.5)

Besides the above mentioned function, this kind machine tools could control the velocity and displacement at each point. It is the most complicated and most flexible way of machine controlling.

Characteristics:

To control two or more axes simultaneously to get desired shape.

To control not only the destinations, but also the paths through which the tool reaches these destinations.

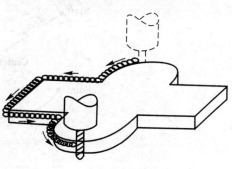

Fig. 10.5

In the process of machining, the tool contacts the workpiece

Interpolation

The method by which contouring machine tools move from one programmed point to the next is called interpolation. This ability to merge individual axis points into a predefined tool path is built into most of today's MCUs. There are five methods of interpolation: linear, circular, helical, parabolic, and cubic. All contouring controls provide linear interpolation, and most controls are capable of both linear and circular interpolation. Helical, parabolic, and cubic interpolation are used by industries that manufacture parts which have complex shapes, such as aerospace parts and dies for car bodies.

Linear Interpolation consists of any programmed points linked together by straight lines, whether the points are close together or far apart (Fig. 10.6). Curves can be produced with linear interpolation by breaking them into short, straight-line segments. This method has limitations, because a very large number of points would have to be programmed to describe the curve in order to produce a contour shape. A contour programmed in linear interpolation requires the coordinate positions (XY positions in two-axis work) for the start and finish of each line segment.

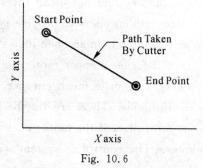

Fig. 10.6

Therefore, the end point of one line or segment becomes the start point for the next segment, and so on, throughout the entire program.

Circular Interpolation: The development of MCUs capable of circular interpolation has greatly simplified the process of programming arcs and circles. To program an arc (Fig.

— 183 —

12), the MCU requires only the coordinate positions (the XY axes) of the circle center, the radius of the circle, the start point and end point of the arc being cut, and the direction in which the arc is to be cut (clockwise or counterclockwise) See Fig. 10.7.

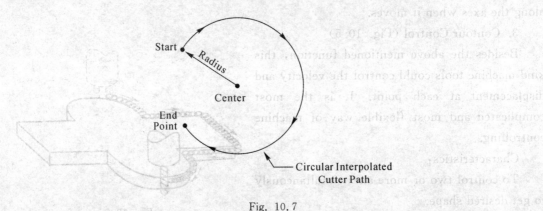

Fig. 10.7

Fig. 10.7: For two-dimensional circular interpolation the MCU must be supplied with the XY axis, radius, start point, end point, and direction of cut.

Classified by the feeding system

A. Open loop refers to a system where the communication between the controller system and the motor is one way (Fig. 10.8). Check the image to the right. As you can see the process for an open loop system is simple. After the user decides what he/she wants to do and generates the G-code or some sort of work file, the NC software then create the necessary step and direction signals to perform the desired task. The computer relays this information to the controller which then energizes the motor/s. After the motor moves to the desired position, there is no feedback to the controller system to verify the action. In the CNC industry, open loop systems use stepper motors. However, just because a system uses stepper motors does not mean the system is an open loop system. Stepper motors may be outfitted with encoders to provide position feedback just like servo motors. Stepper motors are able to operate in an open loop system while servo motors are not, for CNC applications at least. Because stepper motors do not require feedback hardware, the price for an open loop CNC system is much cheaper and simpler than a closed loop system. This makes it more affordable. There are drawbacks to the open loop system. Because there is no feedback to the controller, if the motor does not operate as instructed there is no way for the system to know. The controller system will continue performing the next task as if there is no problem until a limit switch is tripped or the operator resets the machine.

Chapter 10 A Brief Introduction to Modern Machining

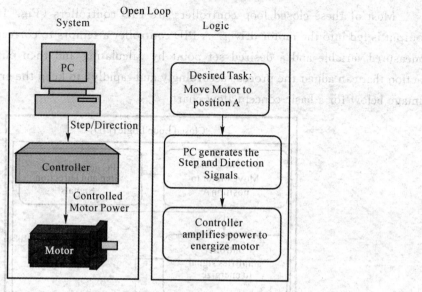

Fig. 10.8

B. The closed loop system has a feedback system to monitor the output of the motors (Fig. 10.9). Closed systems are also able to correct errors in position, velocity, and acceleration, and also fault the system if the error is too large. Refer to the image below. As you can see from the image to the left, there are two closed loop system shown. The first system returns the feedback to the CNC controller. The second system returns the feedback into the computer. Regardless what some say, both systems are true closed loop systems. The system where the feedback is fed into the signal generator or computer is usually found on high end machines.

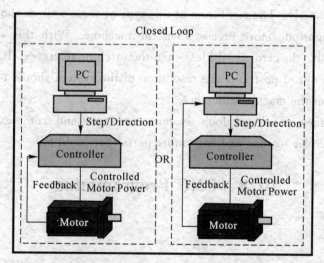

Fig. 10.9

The image on the left represents the most common type of closed loop controller system. In this type of system, an encoder, glass scale, or some other type of analog device is responsible for the feedback signal.

Most of these closed loop controllers are PID controllers (Fig. 10.10). The encoder output is fed into the motor driver. A PID controller attempts to correct the error between a measured variable and a desired set point by calculating and then outputting a corrective action that can adjust the process accordingly and rapidly, to keep the error minimal. See the image below for a basic concept flow chart.

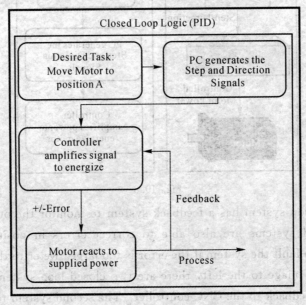

Fig. 10.10

This type of control loop is set to fault at a preset value. This should stop the machine in case of excess error. Some people believe that this type of system can be inaccurate. This is untrue if setup properly. The resolution of this type of servo system should be designed to be one order of magnitude more precise than the machine. With this setup, even if the machine were to fault, the error is still less than the machine tolerance. If a controller faults when it is 124 steps out of position, the resolution of the system should be designed so that 124 steps is less than the machine tolerance.

The disadvantages of closed loop systems are cost and complexity. Closed loop controllers can be harder to tune and have more parts that could fail.

10.8 CNC Programming

Programming Type

1. Manual programming

It is usually used to machine the simple spares whose program is short.

2. Automated programming

The software could form the NC program automatically as long as the drawing has been input.

Basic Concepts

Axis of motiondescribes the relative motion that occurs between the cutting tool and the workpiece. Three main axes of motion for machine tools are referred to as the x, y, and z axes that form a right hand coordinate system (Fig. 10.11).

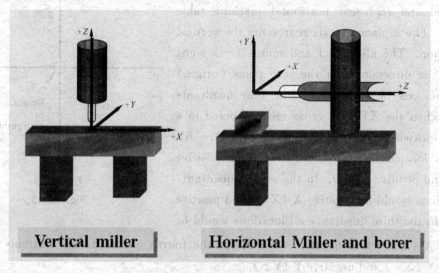

Fig. 10.11

It is not convenient to make NC program with the machine coordinate, sometimes, we choose a point on the work to be the starting point, to set up the coordinate. This coordinate is called the workpart/program coordinate.

Cartesian coordinate system

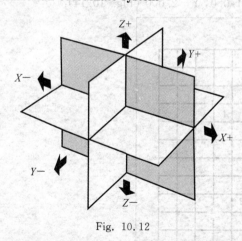

Fig. 10.12

The Cartesian, or rectangular, coordinate system was devised by the French mathematician and philosopher Rene' Descartes. With this system, any specific point can be described in mathematical terms from any other point along three perpendicular axes. This concept fits machine tools perfectly since their construction is generally based on three axes of motion (X, Y, Z) plus an axis of rotation. On a plain vertical milling machine, the X axis is the horizontal movement (right or left) of the table, the Y axis is the table cross movement (toward or away from the column), and the Z axis is the vertical movement of the knee or the spindle. CNC systems rely heavily on the use of rectangular coordinates because the programmer can locate every point on a job precisely. When points are located on

Simplified Metal Works

a work piece, two straight intersecting lines, one vertical and one horizontal, are used. These lines must be at right angles to each other, and the point where they cross is called the origin, or zero point (Fig. 10.12).

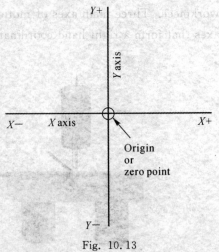

The three-dimensional coordinate planes are shown in left. The X and Y planes (axes) are horizontal and represent horizontal machine table motions. The Z plane or axis represents the vertical tool motion. The plus (+) and minus (−) signs indicate the direction from the zero point (origin) along the axis of movement. The four quadrants formed when the XY axes cross are numbered in a counterclockwise direction (Fig. 10.13). All positions located in quadrant 1 would be positive (X+) and positive (Y+). In the second quadrant, all positions would be negative X (X−) and positive (Y+). In the third quadrant, all locations would be

Fig. 10.13

negative X (X−) and negative (Y−). In the fourth quadrant, all locations would be positive X (X+) and negative Y (Y−).

Below, point A would be 2 units to the right of the Y axis and 2 units above the X axis. Assume that each unit equals 1.000. The location of point A would be X + 2.000 and Y + 2.000. For point B, the location would be X + 1.000 and Y − 2.000. In CNC programming it is not necessary to indicate plus (+) values since these are assumed. However, the minus (−) values must be indicated. For example, the locations of both A and B would be indicated as follows: A X+2.000, Y+2.000; B X+1.000, Y−2.000 (Fig. 10.14).

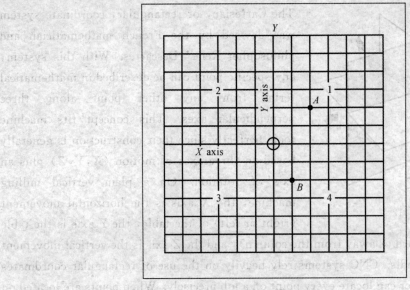

Fig. 10.14

— 188 —

10.8.1 Machine Types

Lathe

The engine lathe, one of the most productive machine tools, has always been an efficient means of producing round parts (Figure below). Most lathes are programmed on two axes. The X axis controls the cross motion of the cutting tool. Negative X ($X-$) moves the tool towards the spindle centerline; positive X moves the tool away from the spindle centerline. The Z axis controls the carriage travel toward or away from the headstock (Fig. 10.15).

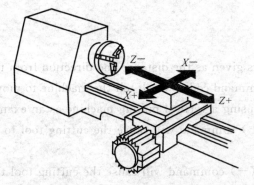

Fig. 10.15 The Main Axes of a Lathe or Turning Center

Milling Machine

The milling machine has always been one of the most versatile machine tools used in industry (Fig. 10.16). Operations such as milling, contouring, gear cutting, drilling, boring, and reaming are only a few of the many operations which can be performed on a milling machine. The milling machine can be programmed on three axes:

• The X axis controls the table movement left or right.

• The Y axis controls the table movement toward or away from the column.

• The Z axis controls the vertical (up or down) movement of the knee or spindle.

10.8.2 Programming Systems

Two types of programming modes, the incremental system and the absolute system, are used for CNC. Both systems have applications in CNC programming, and no system is either right or wrong all the time. Most controls on machine tools today are capable of handling either incremental or absolute

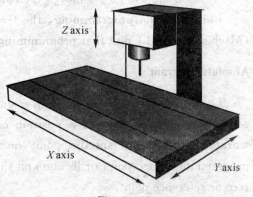

Fig. 10.16

programming (Fig. 10.17).

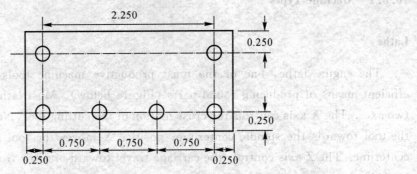

Fig. 10.17 A Workpiece Dimensioned in the Incremental System Mode

Incremental program

Locations are always given as the distance and direction from the immediately preceding point (Fig. 10.16). Command codes which tell the machine to move the table, spindle, and knee are explained here using a vertical milling machine as an example.

- A "X plus" ($X+$) command will cause the cutting tool to be located to the right of the last point.
- A "X minus" ($X-$) command will cause the cutting tool to be located to the left of the last point.
- A "Y plus" ($Y+$) command will cause the cutting tool to be located toward the column.
- A "Y minus" ($Y-$) will cause the cutting tool to be located away from the column.
- A "Z plus" ($Z+$) command will cause the cutting tool or spindle to move up or away from the workpiece.
- A "Z minus" ($Z-$) moves the cutting tool down or into the workpiece.

In incremental programming, the G91 command indicates to the computer and MCU (Machine Control Unit) that programming is in the incremental mode.

Absolute program

Locations are always given from a single fixed zero or origin point (Fig. 10.18). The zero or origin point may be a position on the machine table, such as the corner of the worktable or at any specific point on the workpiece. In absolute dimensioning and programming, each point or location on the workpiece is given as a certain distance from the zero or reference point.

In absolute programming, the G90 command indicates to the computer and MCU that the programming is in the absolute mode.

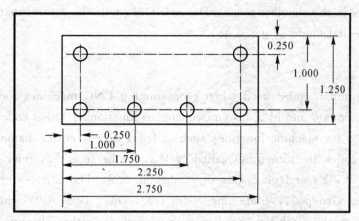

Fig. 10.18 A Workpiece Dimensioned in the Absolute System Mode
Note: All dimensions are given from a known point of reference.

10.8.3 Programming Format

Word address is the most common programming format used for CNC programming systems. This format contains a large number of different codes that transfers program information from the part print to machine servos, relays, micro-switches, etc., to manufacture a part. These codes, which conform to EIA (Electronic Industries Association) standards, are in a logical sequence called a block of information. Each block should contain enough information to perform one machining operation.

Word Address Format

Every program for any part to be machined must be put in a format that the machine control unit can understand. The format used on any CNC machine is built in by the machine tool builder and is based on the type of control unit on the machine. A variable-block format which uses words (letters) is most commonly used. Each instruction word consists of an address character, such as X, Y, Z, G, M, or S. Numerical data follows this address character to identify a specific function such as the distance, feed rate, or speed value. The address code G90 in a program, tells the control that all

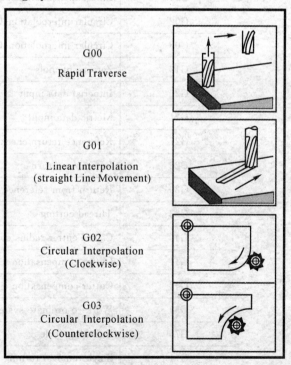

Fig. 10.19

measurements are in the absolute mode. The code G91, tells the control that measurements are in the incremental mode (Fig. 10.19).

Coding

The most common codes used when programming CNC machines tools are G-codes (preparatory functions), and M codes (miscellaneous functions). Other codes such as F, S, D, and T are used for machine functions such as feed, speed, cutter diameter offset, tool number, etc. G-codes are sometimes called cycle codes because they refer to some action occurring on the X, Y, and/or Z axis of a machine tool. The G-codes are grouped into categories such as Group 01, containing codes G00, G01, G02, G03 which cause some movement of the machine table or head. Group 03 includes either absolute or incremental programming, while Group 09 deals with canned cycles. A G00 code rapidly positions the cutting tool while it is above the workpiece from one point to another point on a job. During the rapid traverse movement, either the X or Y axis can be moved individually or both axes can be moved at the same time. Although the rate of rapid travel varies from machine to machine, it ranges between (5 and 20 m/min).

Talbe 10.1 Some of the Most Common G-codes Used in CNC Programming.

G-CODE	DESCRIPTION
G00	Rapidtraverse (linear)
G01	Linear interpolationat feedrate
G02	Circular interpolation(CW) R+
G03	Circular interpolation(CCW) R+
G10	Zero point tools
G20	Imperial data input
G21	Metric data input
G27	Reference return check
G28	Reference return
G29	Return from reference
G33	Thread cutting
G40	Cancel cutter radius compensation
G41	Cutter compensation left
G42	Cutter compensation right
G54	Work co-ordinate system no. 1
G70	Finishing cycle
G71	Stock removal turning

(Continued)

G-CODE	DESCRIPTION
G72	Stock removal facing
G74	Face grooving
G75	Diameter grooving
G76	Threading cycle
G90	Absolute date input
G91	Incremental data input
G94	Feed /minute mode
G95	Feed /rev mode
G96	Constant surface speed mode
G97	Cancel constant surface speed

G-code is a common name for the programming language that is used for NC and CNC machine tools. It is defined in EIA RS-274-D. G-code is also the name of any word in a CNC program that begins with the letter G, and generally is a code telling the machine tool what type of action to perform, such as:
- rapid move
- controlled feed move in straight line or arc
- Series of controlled feed moves that would result in a hole being drilled.
- change a pallet
- Set tool information such as offset.

There are other codes; the type codes can be thought of like registers in a computer
- X position
- Y position
- Z position
- M code (another "action" register)
- F feed rate
- S spindle speed
- N line number
- R Radius
- T Tool selection
- I Arc data X axis
- J Arc data Y axis.

M or miscellaneous codes are used to either turn ON or OFF different functions which control certain machine tool operations. M-codes are not grouped into categories, although several codes may control the same type of operations such as M03, M04, and M05 which

control the machine tool spindle.
- M03 turns the spindle on clockwise
- M04 turns the spindle on counterclockwise
- M05 turns the spindle off.

Table 10.2　Some of the Most Common M-codes Used in CNC Programming.

Code	Function
M00	Program stop
M02	End of program
M03	Spindle start (forward CW)
M04	Spindle start (reverse CCW)
M05	Spindle stop
M06	Tool change
M08	Coolant on
M09	Coolant off
M10	Chuck-clamping(＊＊)
M11	Chuck-unclamping(＊＊)
M12	Tailstock spindle out (＊＊)
M13	Tailstock spindle in(＊＊)
M17	Toolpost rotation normal(＊＊)
M18	Toolpost rotation reverse(＊＊)
M30	End of tape and rewind
M98	Transfer to subprogram
M99	End of subprogram
(＊＊)	Refers only to CNC lathes and turning centre

Block of Information

CNC information is generally programmed in blocks of five words. Each word conforms to the EIA standards and they are written on a horizontal line. If five complete words are not included in each block, the machine control unit (MCU) will not recognize the information, therefore the control unit will not be activated. Using the example shown in below, the five words are as follows: N001 represents the sequence number of the operation. G01 represents linear interpolation X12345 will move the table 1.2345 in. in a positive direction along the X axis. Y06789 will move the table 0.6789 in. along the Y axis. M03 Spindle on CW (Fig. 10.20).

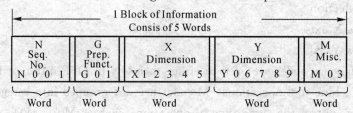

Fig. 10.20　A Complete Block of Information Consists of Five Words

Chapter 10 A Brief Introduction to Modern Machining

Programming for Positioning

Before starting to program a job, it is important to become familiar with the part to be produced. From the engineering drawings, the programmer should be capable of planning the machining sequences required to produce the part. Visual concepts must be put into a written manuscript as the first step in developing a part program, see Fig. 10.21. It is the part program that will be sent to the machine control unit by the computer, tape, diskette, or other input media. The programmer must first establish a reference point for aligning the workpiece and the machine tool for programming purposes. The manuscript must include this along with the types of cutting tools and work-holding devices required, and where they are to be located.

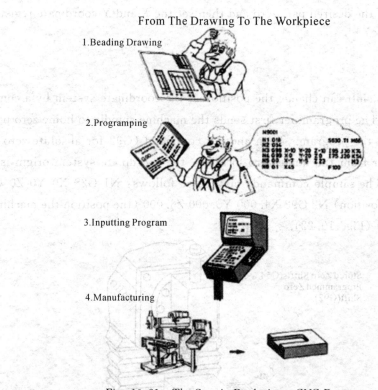

Fig. 10.21 The Step in Producing a CNC Program

Dimensioning Guidelines

The system of rectangular coordinates is very important to the successful operation of CNC machines. Certain guidelines should be observed when dimensioning parts for CNC machining. The following guidelines will insure that the dimensioning language means exactly the same thing to the design engineer, the technician, the programmer, and the machine operator.

— 195 —

(1) Define part surfaces from three perpendicular reference planes.

(2) Establish reference planes along part surfaces which are parallel to the machine axes.

(3) Dimension from a specific point on the part surface.

(4) Dimension the part clearly so that its shape can be understood without making mathematical calculations or guesses.

(5) Define the part so that a computer numerical control cutter path can be easily programmed.

Machine Zero Point

The machine zero point can be set by three methods — by the operator, manually by a programmed absolute zero shift, or by work coordinates, to suit the holding fixture or the part to be machined. MANUAL SETTING - The operator can use the MCU controls to locate the spindle over the desired part zero and then set the X and Y coordinate registers on the console to zero.

Absolute Zero Shift

The absolute zero shift can change the position of the coordinate system by a command in the CNC program. The programmer first sends the machine spindle to home zero position by a G28 command in the program. Then another command (G92 for absolute zero shift) tells the MCU how far from the home zero location, the coordinate system origin is to be positioned, Fig. 13. The sample commands may be as follows: N1 G28 X0 Y0 Z0 (sends spindle to home zero position) N2 G92 X4.000 Y5.000 Z6.000 (the position the machine will reference as part zero) (Fig. 10.22).

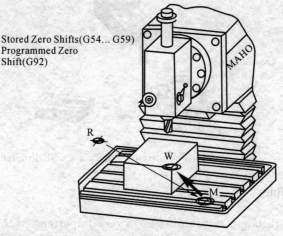

R=Reference point (maximum travel of machine)

M=Machine zero point(X0.Y0.Z0)of machine coordinate system.

W=Part zero point workpiece coordinate system.

Fig. 10.22 The Relationship Between the Part Zero and the Machine System of Coordinates.

Chapter 10 A Brief Introduction to Modern Machining

Work Settings and Offsets

All CNC machine tools require some form of work setting, tool setting, and offsets (compensation) to place the cutter and work in the proper relationship. Compensation allows the programmer to make adjustments for unexpected tooling and setup conditions.

Work Coordinates

In absolute positioning, work coordinates are generally set on one edge or corner of a part and all programming is generally taken from this position. In Fig. 10.23, the part zero is used for all positioning for hole locations 1, 2, and 3.

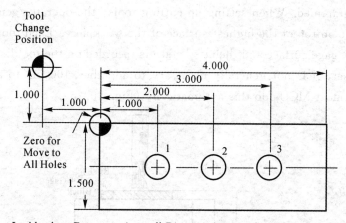

Fig. 10.23 In Absolute Programming, all Dimensions must be Taken from the XY Zero at the Top Left-hand Corner of the Part

In incremental positioning, the work coordinates change because each location is the zero point for the move to the next location (Fig. 10.24).

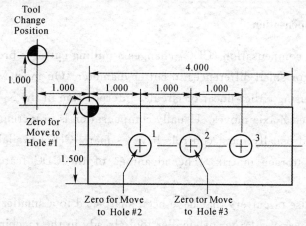

Fig. 10.24 In Incremental Programming, all Dimensions are Taken from the Previous Point.

On some parts, it may be desirable to change from absolute to incremental, or vice versa, at certain points in the job. Inserting the G90 (absolute) or the G91 (incremental) command into the program at the point where the change is to be made can do this.

R Plane or Gage Height

The letter R refers to a partial retraction point in the Z axis to which the end of the cutter retracts above the work surface to allow safe table movement in the X Y axes. It is often called the rapid-traverse distance, gage height, retract or work plane. The R distance is a specific height or distance above the work surface and is generally .100 in. above the highest surface of the workpiece, Figure below, which is also known as gage height. Some manufacturers build a gage height distance of .100 in. into the MCU (machine control unit) and whenever the feed motion in the Z axis is called for, .100 in. will automatically be added to the depth programmed. When setting up cutting tools, the operator generally places a .100 in. thick gage on top of the highest surface of the workpiece. Each tool is lowered until it just touches the gage surface and then its length is recorded on the tool list. Once the gage height has been set, it is not generally necessary to add the .100 in. to any future depth dimensions since most MCUs do this automatically (Fig. 10.25).

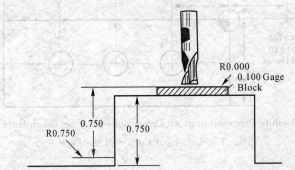

Fig. 10.25 Using a .100 in. Gage Block to Set the Gage Height or R0 on the Work Surface

Cutter Diameter Compensation

Cutter diameter compensation (CDC) changes a milling cutter's programmed centerline path to compensate for small differences in cutter diameter. On most MCUs, it is effective for most cuts made using either linear or circular interpolation in the X-Y axis, but does not affect the programmed Z-axis moves. Usually compensation is in increments of .0001 in. up to +1.0000 in., and usually most controls have as many CDCs available as there are tool pockets in the tool storage matrix. The advantage of the CDC feature is that it (Fig. 10.26):

(1) Allows the use of cutters that have been sharpened to a smaller diameter.

(2) Permits the use of a larger or smaller tool already in the machine's storage matrix.

(3) Allows backing the tool away when roughing cuts are required due to excessive material present.

(4) Permits compensation for unexpected tool or part deflection, if the deflection is

constant throughout the programmed path. The basic reference point of the machine tool is never at the cutting edge of a milling cutter, but at some point on its periphery. If a 1.000 in. diameter end mill is used to machine the edges of a workpiece, the programmer would have to keep a .500 in. offset from the work surface in order to cut the edges accurately, Figure below. The .500 offset represents the distance from the centerline of the cutter or machine spindle to the edge of the part. Whenever a part is being machined, the programmer must calculate an offset path, which is usually half the cutter diameter.

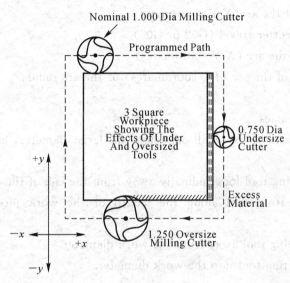

Fig. 10.26 Cutter-diameter Compensation must be Used When Machining with Various Size Cutters

Modern MCUs, which have part surface programming, automatically calculate centerline offsets once the diameter of the cutter for each operation is programmed. Many MCUs have operator-entry capabilities which can compensate for differences in cutter diameters; therefore an oversize cutter, or one that has been sharpened, can be used as long as the compensation value for oversize or undersize cutters is entered.

CNC Programming Hints — MILLING

⊕ Machine reference point (maximum travel of machine)
⊕ Machine XY zero point (could be tool change point)
⊕ Part XY zero point (programming start point)
✢ Indicates the tool change position. A G92 code will reset the axis register position coordinates to this position.

For program to run on a machine, it must contain the following codes:

M03 to start the spindle/cutter revolving

Sxxx The spindle speed code to set the r/min.

Fxx The feed rate code to move the cutting tool or workpiece to the desired position.

Angles: The XY coordinates of the start point and end point of the angular surface plus a feed rate (F) are required.

Z codes:

A Z dimension raises the cutter above the work surface.

A Z- dimension feeds the cutter into the work surface.

Z. 100 is the recommended retract distance above the work surface before a rapid move (G00) is made to another location.

RADII/CONTOUR Requirements:

The start point of the arc (X Y coordinates)

The direction of cutter travel (G02 or G03)

The end point of the arc (X Y coordinates)

The centre point of the arc (I J coordinates) or the arc radius.

Tapers/Bevels/Angles

The X Z coordinates of the small diameter, the large diameter, and a federate must be programmed.

Z moves the cutting tool longitudinally away from the end of the workpiece

Z- moves the cutting tool along the length of the work piece toward the chuck (headstock)

X moves the cutting tool away from the work diameter.

X- moves the cutting tool into the work diameter.

Other Codes

Table 10.3

Function	Code	Meaning and Examples
Part Code	O	O1-9999
Code Section	N	N1-9999
Motion Path	G	linear, circle, G00-G99
Motion Direction and Distance	XYZ ABC UVW	Move in the axis -99999.99-_99999.99
Radius	R	Radius
Circus Centre Coordinate	I J K	Distance between circus center and the starting coordinate
Feeding Speed	F	F0-15000
Spindle Speed	S	S0-9999
Tool Selection	T	T0-99
Auxiliary Switches	M	M0-99
Offset	HD	00-99
Pause	P	

(Continued)

Function	Code	Meaning and Examples
Program Code Set	P	P1-9999
Repetition	L	1,2,3…
Parameter	P Q R	Parameter for cycling

Examples 1 (Fig. 10.27)

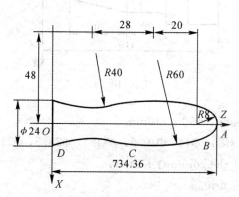

Fig. 10.27

Suppose the O point is the program zero, A (0, 73.436), B (7.385, 68.513), C (9.6, 28.636), D (12, 0)

%0010;   define the program number
N01 G92 X16 Z73.436;   setting up the workpiece coordinates (by tool position)
N02 M03 S500;
N03 M98 P0006 L11;   run subprogram 0006, 11 times, finish the cutting
N04 M05;   spindle stop
N05 M02;   end of program
%0006   subprogram 0006
N02 G91 G01 X-6.0 F500   cutting off ($Z=73.436$) along the x- direction.
N03 G03 X7.385 Z-4.923 R8; process the arc A_iB_i
N04   X2.215 Z-28.636 R60; process the arc B_iC_i
N05 G02 X2.4 Z-28.636 R40; Process the arc C_iD_i
N06 G00 X2.0;   cutting along the X($Z=0$)
N07 Z73.436;   cutting along the Z
N08 X-9.0;   cuttingalong the x-
N09 M99;   return the subprogram

Example 2 (Fig. 10. 28)

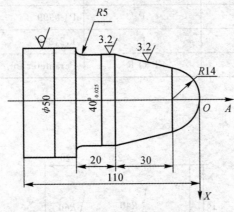

Fig. 10. 28

The workpiece dimension: Φ50 * 110mm

Right side processing step: Φ44, Φ40.5, Φ34.5, Φ28.5, Φ22.5, Φ16.5

Outer curve surface of the column: R14.25

Outer column surface: Φ40.5

Outer cone curve surface: R4.75

Fine cutting: curve outer surface R14

Fine cutting: outer column surface: Φ40

Fine cutting: R5

Main spindle speed: 630r/min

Feeding speed: rough processing 0.2mm/r, fine processing 0.1mm/r

Size: R14 curve center point: $x=0$, $z=-14$ mm

R5 curve center point: $x=50$mm, $z=-(44+20-5)$mm$=-59$mm

Reference coding

The incremental quantity will marked by U, W

N10 G50 X100.0 Z100.0; setting up the workpiece coordinates

N20 S630 M03 T11; main spindle start n=630 r/min, tool 1, compensation 1.

N30 G00 X52.0 F0.2; fast positioning

N40 G01 X0.0 F0.2; processing the right end

N50 G00 Z1.0; fast positioning

N60 X44.0;

N70 G01 Z-62.5; rough processing outer column Φ44

N80 X50.0; processing the step

N90 G00 Z1.0; fast positioning

N100 X40.5

N110 G01 Z-60.0; rough processing the outer column to 40.5
N120 X44.0; rough processing the step
N130 G00 Z1.0; fast positioning
N140 X34.5;
N150 G01 Z-29.0; rough processing the outer column to $\Phi 40.5$
N160 X40.5;
N170 G00 Z1.0; fast positioning
N180 X28.5;
N190 G01 Z-14.0;
N200 X34.5; processing the step
N210 G00 Z1.0; fast positioning
N220 X22.5;
N230 G01 Z-4.0; rough processing the outer column to $\Phi 22.5$
N240 X28.5;
N250 G00 Z1.0; fast positioning
N260 X16.5;
N270 G01 Z-2.0; rough processing the outer column to $\Phi 16.5$
N280 X22.5; processing the step
N290 G00 Z0.25; fast positioning
N300 X0.0;
N310 G03 X28.5 Z-14.0 R14.25; Rough processing the curve with R14.25
N320 G01 X40.5 Z-44.0; rough processing the cone
N330 W-15.0; rough processing the column to $\Phi 40.5$
N340 G02 X50.0 W-4075 R4.75; rough processing the curve with R4.75
N350 G00 Z0.0; fast positioning
N360 X0.0;
N370 G03 X28.0 Z-14.0 R14.0; fine processing the curve with R14
N380 G01 X40.0 Z-44.0; fine processing outer curve
N390 W-15.0; fine processing outer curve with $\Phi 40$
N400 G02 X50.0 W-5.0 R5.0; fine processing the curve R5
N410 G00 X100.0 Z100.0 T10.0; fast position the tool to the original point, remove the tool compensation
N420 M05; stop the main spindle
N430 M30; stop the program

There are also programming habit for CNC milling machine and Machining Centre, but for a brief introduction, they are way beyond this material.

Exercises

1. Complete the following table.

Table 10.4

G-Code	Function Description
G00	
G01	
G02	
G03	
G10	
G20	
G21	
G27	
G28	
G29	
G33	
G40	
G41	
G42	
G54	
G70	
G71	
G72	
G74	
G75	
G76	
G90	
G91	
G94	
G95	
G96	
G97	

2. Complete the following table.

Table 10.5

M-Code	Function
M00	
M02	
M03	
M04	
M05	
M06	
M08	
M09	
M10	
M11	
M12	
M13	
M17	
M18	
M30	
M98	
M99	
(**)